SpringerBriefs in Systems Biology

For further volumes:
http://www.springer.com/series/10426

Vijaykumar Yogesh Muley
Vishal Acharya

Genome-Wide Prediction and Analysis of Protein–Protein Functional Linkages in Bacteria

Springer

Vijaykumar Yogesh Muley
Center of Excellence in Epigenetics
Indian Institute of Science
Education and Research (IISER)
Pune, Maharashtra
India

Vishal Acharya
Biotechnology Division
CSIR-Institute of Himalayan Bioresource
Technology (IHBT)
Palampur, Himachal Pradesh
India

ISSN 2193-4746 ISSN 2193-4754 (electronic)
ISBN 978-1-4614-4704-7 ISBN 978-1-4614-4705-4 (eBook)
DOI 10.1007/978-1-4614-4705-4
Springer New York Heidelberg Dordrecht London

Library of Congress Control Number: 2012942706

Printed on acid-free paper

Springer is part of Springer Science+Business Media (www.springer.com)

"What I cannot create, I do not understand"
-Richard Feynman

Dedicated to my Parents, my niece Neha, nephew Pravin and Avinash.

Acknowledgments

It is my immense pleasure to utilize this space to acknowledge all those who directly or indirectly helped me in the accomplishment of this work. First and foremost, I would like to extend gratitude to my Ph.D. mentor Dr. Akash Ranjan, Centre for DNA Fingerprinting and Diagnostics, Hyderabad, India. He has helped me evolve in the scientific field through the Darwinian principle, "survival of the fittest". I also like to thank, Dr. Kapil Kamble, Department of Microbiology, Sant Gadge Baba Amravati University, Amravati, India for helpful discussion and sharing inspirational thoughts on protein interactions. I would also like to thank my former lab members Manjari, Anupam Sinha and Subbaiah (CDFD, Hyderabad) for scintillating discussions on post-genomic solutions using computations methods. This part of the work is supported by grant from Centre of Excellence in Epigenetics program of the Department of Biotechnology, Government of India, India. My sincere thank to Prof. Sanjeev Galande for supporting me by the Swarnajayanti fellowship to him at IISER, Pune. Vishal Acharya would like to thank Paramvir Singh Ahuja, Director, CSIR-Institute of Himalayan Bioresource Technology, Palampur, India.

Finally, Melanie Tucker and Meredith Clinton, Springers editorial members, should be specially acknowledged for their patience during the preparation of this book.

Vijaykumar Muley

Acknowledgments

Contents

Chapter 1
Introduction

Abstract The cell is a crowded laboratory where active synthesis and degradation of various molecules is a routine protocol. Signals in response to internal or external perturbations have to be channeled through these molecules in order to invoke appropriate cellular response. Therefore, the functioning of cellular systems totally dependent on interactions between inter- and intramolecular components. Public availability of several completely sequenced genomes at the end of the twentieth century has provided the necessary platforms to elucidate these biomolecular interactions on a large scale. These interactions are conceptualized in the form of networks or graphs. As a result, the last decade has witnessed significant progress in the understanding of organization of cellular processes at systems level. In this book, we provide an overview on protein function, prediction of protein–protein interactions, network analyses, and touch upon significance of these studies.

The cell is a dense crowd of molecular and macromolecular components. Proteins are special macromolecules involved in regulation and are responsible for the turnover of a majority of molecular components. Several molecular principles underlying functioning of cell as a unit have emerged from comparative analysis of genes and their products (proteins) in completely sequenced genomes. The most notable was the realization that reductionist approaches for the analysis of gene functions are not enough to understand the complexity of cellular systems. This is because the cell as a whole is organized in the form of various genetic modules resulting from different selective and functional constraints during evolution. Each genetic module often encodes for proteins that interact with each other to perform specific biological tasks such as protein biosynthesis, motility, etc. The expression of these genetic modules is amenable to modulate by metabolites and many small molecules acting as signals. Although these genetic modules play independent functional roles in a cell, the cellular response to various environmental signals is often driven by the cross-talk between them. Therefore, every molecule and macromolecule in the cell is connected with each other either directly or indirectly

V. Y. Muley and V. Acharya, *Genome-Wide Prediction and Analysis of Protein–Protein Functional Linkages in Bacteria*, SpringerBriefs in Systems Biology, DOI: 10.1007/978-1-4614-4705-4_1,

via adjacent molecules (neighbors). However, organization, evolution, and functional constraints underlying various biological processes at system level were largely unknown. Several completely sequenced genomes available since the late 1990s have provided essential ingredients to address these issues, whereas the technical background stemmed from the computational and mathematical sciences for such studies. The synergy between biological and other scientific disciplines led to chart possible interactions among biological components. The dissection of these interactions revealed that molecules in living organisms are highly connected and show intrinsic modularity. The handy way to represent such complex data is a network or graph where molecules are nodes and connecting links between them are edges. Fortunately, the network theory also emerged as contemporary to dissect these complex networks, which were already making huge impact on the physical sciences during this period. This resulted in networks becoming an important part of molecular and cellular biology.

In light of this development, in the subsequent chapters, we describe protein functions and on top of it build the concept of protein–protein interaction networks. Furthermore, we discuss some of the important methods that utilize genomic information to predict physical and functional protein–protein interactions.

Chapter 2
From Genomes to Protein Functions

Abstract Prokaryotic organisms are constantly challenged by fluctuations in the surrounding environment. Due to unicellular nature, the only barrier that separates prokaryotic cell from the environment is a thin layer of cell membrane. In the course of evolution, prokaryotic organisms have acquired numerous phenotypes, diverse metabolic activities, and more importantly, sensing systems to overcome the environmental perturbations and thereby maintains its structural integrity even in the extreme surrounding conditions. This battle between cell and environment is constantly monitored and played by fine-tuning of micro- and macromolecular constituents of the cell. Proteins are one of the important dynamic macromolecules, which practically regulate and control these orchestras and are virtually involved in every aspect of the cellular activities. This chapter describes with multi-faceted functional roles of proteins, which allow them to perform diverse activities. Eventually, we also provide post-genomic resources and systematic tools useful for understanding protein's function with minimal efforts.

2.1 Introduction

The fate of every living organism is written on its genome sequence in the form of arrays of genes. The expression of these genes is responsible for the organism's observable features, commonly known as phenotypes or traits, such as its morphology, development, behavior, and biochemical properties. The nutrient availability and various environmental factors influence the expression of genes. As a result, the organism expresses only a subset of its total genes crucial for survival at a particular point of time by taking into account the surrounding conditions. Therefore, the composite effect of environmental components, products of expressed genes, and the interaction between both factors determine the specific phenotype of an

V. Y. Muley and V. Acharya, *Genome-Wide Prediction and Analysis of Protein–Protein Functional Linkages in Bacteria*, SpringerBriefs in Systems Biology, DOI: 10.1007/978-1-4614-4705-4_2,

organism. Proteins are one of the gene products involved in almost every physiological function of an organism. Researchers from various corners of the globe are trying to understand various functional aspects of proteins in model organisms.

Prokaryotic organisms have served as best models for studying protein functions due to their simple unicellular organization and their existence on Earth for longer periods than any other living organisms. Thereby, the protein function related studies in these organisms often provide a key to the evolutionary scale. Moreover, prokaryotes are the most abundant living forms and they have conquered every bio-geographical space on the Earth. They are omnipresent, from hydrothermal vents where temperature is more than 80 °C, associated with the host like human, sewage treatment plants, nearby hazardous pollutants, coal mines to that of frigid atmosphere of Antarctica. Therefore, identification of proteins responsible for adaptation to diverse ecological niche may lead to the discovery of novel biochemical and cellular processes to degrade hazardous pollutants, to generate sustainable energy, and to understand the molecular basis of diseases which are manifestation of complex phenotypic phenomena.

The public available completely sequenced genomes are making it possible to grow in-depth knowledge on the genomic organization and the architecture of proteins to define their biological functions at different scales. One of the major goals of post-genomic era is to study functions of protein repertoire of an organism for understanding, how tiny unicellular bacteria have evolved and diversified into various complex multicellular life-forms with astonishing diverse phenotypes we observe today. In this chapter, we focus on basics of protein functions with respect to computational analysis and eventually build reasons for studying them in the context of others.

2.2 Proteins in the Post-Genomic Era

Proteins are composed of covalently connected amino acid units. Each protein has its own unique amino acid sequence encoded by the nucleotide sequence of the respective gene. Protein participates in virtually every aspect of biological process within the cell. Three important aspects are associated with functions of protein that include its subcellular localization and its role at the molecular and the cellular levels [1–5]. For instance, at the cellular level, isoforms of hexokinase belongs to the different subcellular localization catalyze reaction in the glycolysis pathway to generate energy. The molecular function of a hexokinase is in phosphorylation of the Hydroxyl group (-OH) of a six-carbon sugar substrate, which is one of the precursors for energy generation via glycolysis pathway. Conventionally, protein's biological role has been elucidated by various experimental methods and often its three-dimensional structure (3-D) has been used as a framework to explain known functional properties.

With the advance in sequencing technology, genomic information databases have been flooded with the sequences. As of September 2011, the Genomes

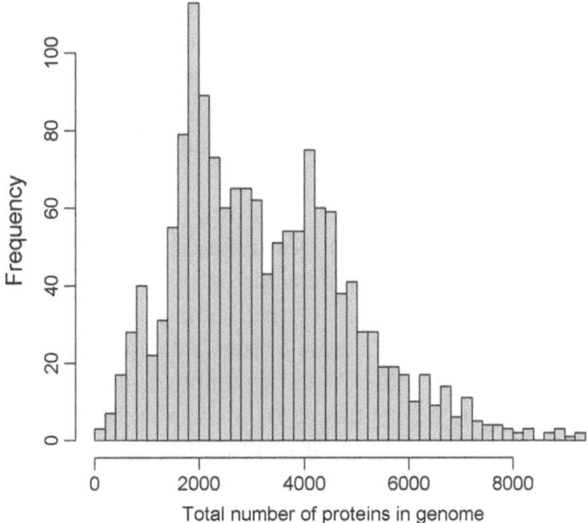

Fig. 2.1 A frequency distribution of total numbers of proteins encoded by 1,491 prokaryotes

OnLine Database (GOLD) contained information for 2,907 complete genome sequences out of 11,472 ongoing sequencing projects [6]. The frequency distribution of numbers of proteins encoded by 1,491 prokaryotic genomes suggests substantial differences (Fig. 2.1). The total number of proteins encoded by these genomes range from 121 of endosymbiont *Candidatus Tremblaya princeps* to 9,381 of soil-dwelling *Sorangium cellulosum* bacteria. *Sorangium cellulosum* is stands out as largest prokaryotic genome sequenced till date [7], whereas *Candidatus Tremblaya princeps* is so small that it cannot be considered an independent organism [8]. Majority of prokaryotes have capacity to encode 1,200–6,000 proteins. This huge variation in the number of proteins encoded by prokaryotes is most probably a reflection of their exhaustive lifestyles and diversity of functions.

Several cutting-edge computational approaches have been developed to ease the functional analysis of predicted proteins in the newly sequenced genomes [9–13]. Majority of these approaches use sequences or structural information of well-characterized, existing proteins as templates to infer function for unknown proteins. The rationale behind such approaches is that the high level of sequence or structural similarity of unknown protein to the characterized proteins is likely a common origin (or ancestral relationship); thereby, they should perform similar functional role if not identical. Classically, these approaches are referred as homology-based methods. The term *homology* of proteins is referred when they are derived from common ancestor and they are known as homologous proteins [14]. Homology-based protein function prediction involves the following steps,

1. Search protein of your interest (often of unknown function) against the database of experimentally characterized proteins using Basic Local Alignment Search Tool (BLAST).

2. If BLAST search detects similar sequences for query protein with e-value scores less than 1e-4, one can assume the presence of query protein's homologs in the given database. Homologous proteins can be termed either *orthologs* or *paralogs* that are derived from common ancestor [14–16]. The term ortholog is used to describe proteins that are a result of speciation event and perform analogous functions in different organisms. The term paralog is used to describe homologous proteins which are a result of duplication event and they may perform similar functions in the same organism. Therefore, homology-based protein function prediction depends on orthologous relationship of query protein to its experimentally characterized homolog.
3. In order to detect these events, we can construct Multiple Sequence Alignment (MSA) using one of the available tools. For example, ClustalW, Muscle, etc. [13, 17]. Many times, it is possible to refine MSA positions manually for functionally important residues based on some prior knowledge.
4. The MSA can then be used to construct phylogenetic tree. Phylogenetic tree provides evolutionary trajectory of the proteins.
5. If the branch of tree where our protein of interest falls also includes experimentally characterized proteins, then its probable function is more likely to be similar to them.
6. Figure 2.2 demonstrates how phylogenetic tree is used to deduce functional role of uncharacterized protein.

In spite of these efforts, experimentally characterized protein set is very small and we can transfer annotation (putative functions) to additional but limited number of proteins that have close sequence or structure similarity to well-characterized proteins. The remaining uncharacterized proteins are still named as "hypothetical or unknown". Fraction of such uncharacterized proteins is quite high in the completely sequenced genomes that have been annotated mainly using homology-based methods. Figure 2.3a shows the percentages of hypothetical proteins encoded in 1,491 prokaryotic genomes. On an average 32 % of the proteins encoded by these organisms are referred to as uncharacterized or hypothetical or proteins with unknown functions (Fig. 2.3a). In fact, the percentage of hypothetical proteins goes above 60 in many bacterial genomes. Some of them are medically important pathogenic organisms, such as tuberculosis causing strain of *Mycobacterium avium*, typhoid causing strain of *Salmonella enterica*, and hemolytic species of *Staphylococcus* genus. We plotted the percentages of hypothetical proteins as a function of total numbers of proteins encoded by 1,491 organisms (Fig. 2.3b). There is no correlation between the percentages of hypothetical proteins and the total number of proteins encoded by these genomes. It means that even genomes with smaller number of proteins have also not been well annotated using the available homology-based approaches. These data suggests our inability to unravel the functional roles of several proteins encoded in a genome using homology-based approaches (Fig. 2.3).

Recently, several online services developed for homology-based function predictions and some of them are listed in Table 2.1.

Fig. 2.2 A phylogenetic tree of *Escherichia coli* MioC protein orthologs from various prokaryotic genomes. MioC is a flavoprotein that is necessary for activity of biotin synthesis enzyme. Biotin is a water-soluble B-complex vitamin. *Black* color taxa represent manually curated protein sequences. *Gray* color taxa represent sequences that are very similar to MioC proteins but neither characterized as MioC nor annotated in NCBI database. The phylogenetic tree suggests all these proteins belong to the same family and hence probably perform same functional role

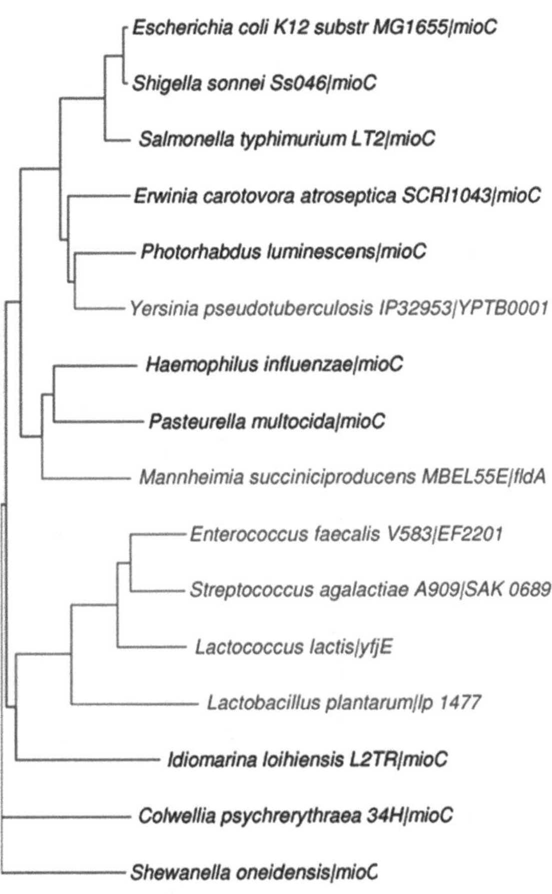

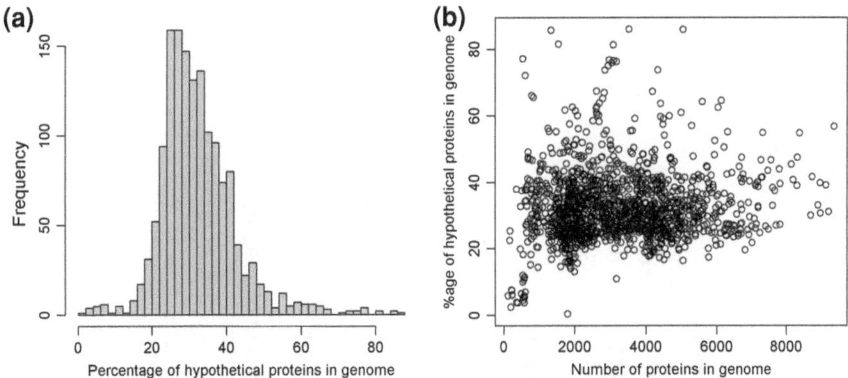

Fig. 2.3 Distribution of hypothetical proteins in the completely sequenced genomes. **a** The frequency distribution of hypothetical proteins, **b** percentage of hypothetical proteins in genomes as a function of total number of proteins. Most of the prokaryotes are not well studied and percentage of hypothetical proteins in them is in between 20 and 40 %

Table 2.1 A list of Web servers dedicated for functional annotation of proteins

Type of protein function	Name of database/tool	Dedicated web server	Remark
Metabolic pathways	KEGG	http://www.genome.jp/kegg/	Pathway database
	BioCyc	http://biocyc.org/	Pathway database
Molecular and biological function	FFPRED	http://bioinf.cs.ucl.ac.uk/ffpred/	Stand-alone, homology-based, multi -purpose
	ESG	http://kiharalab.org/web/esg.php/	Homology-based
	CombFunc	http://www.sbg.bio.ic.ac.uk/ ~ mwass/combfunc/	Homology-based, multi-purpose
Molecular function	BLAST	http://blast.ncbi.nlm.nih.gov/	Homology-based
	SIFTER	http://sifter.berkeley.edu/	Stand-alone, homology-based, phylogeny-assisted function prediction
Pathway/biological function	COG	http://www.ncbi.nlm.nih.gov/COG/	Orthologous groups
Signal transduction	MIST2.1	http://mistdb.com/	Signaling proteins from completely sequence genomes
	SignalCensus	http://www.ncbi.nlm.nih.gov/Complete_Genomes/SignalCensus.html	Signaling proteins from completely sequence genomes
Sub-cellular localization	Phobius	http://phobius.sbc.su.se/	Signal peptide and transmembrane protein analysis
	PSORTb v3.0	http://www.psort.org/psortb/index.html	Multiple subcellular localization prediction
	Octopus	http://octopus.cbr.su.se/index.php	Transmembrane proteins
Transcription factor prediction	DBD	http://dbd.mrc-lmb.cam.ac.uk/DBD/index.cgi?Home	A database of predicted transcription factors in completely sequenced genomes
Transcription regulation	RegulonDB	http://regulondb.ccg.unam.mx/	Escherichia coli-specific database

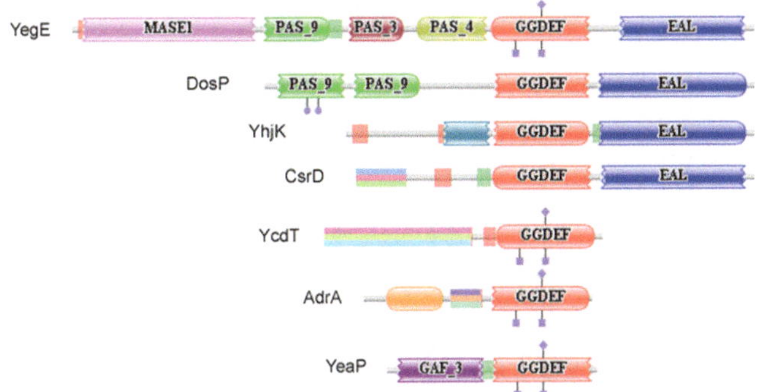

Fig. 2.4 The architecture of GGDEF domain containing seven proteins of *Escherichia coli.* GGDEF is responsible for synthesis of secondary messenger which modulates on environmental cues. The GGDEF domain containing proteins is found in multiple numbers in various bacteria. The figure shows new domains have been added at the N-terminus of the sequences. Top two proteins also show duplication of PAS domain which known to be involved in oxygen sensing. Although every protein has GGDEF domain, their functional role is different due to the combination of other domains

2.2.1 Protein Classification and Architecture

It has been known for a long time that sequences of many proteins can be divided into independent evolutionary units. These independent evolutionary units form structural part of proteins called domains that fold independently and perform unique functions. In the last two decades, analysis of protein domains in completely sequenced genomes has provided us with functional and evolutionary insights of proteins [18–22]. As shown in Fig. 2.4, domains can duplicate, rearrange, and combine in different ways to give rise to new proteins, and hence lead to the functional diversity using same sets of domains. The arrangement of domains on the proteins is its architecture. Proteins are products of genes, and at genome level, various protein architectures are results of duplication, recombination, fusion, and fission of genes.

We have plotted percentage of proteins with multi-domain architecture, multiple copies of the same domain (i.e., Multi-single), and the proteins with single domains. As shown in Fig. 2.5, approximately 30 % of proteins are multi-domains. A small fraction of domain has been duplicated in same proteins (Fig. 2.5). Majority of these proteins perform regulatory activities and the various domain combinations provide versatility in the processes such as signaling, transcription regulation, etc. Since domain is an independent evolutionary unit and forms basic unit of protein organization, it has been used to classify proteins at three hierarchical levels. At the lowest level, domains are classified into "families" based on significant sequence similarity and functional relatedness. These families are grouped into "superfamilies" based on their common origin determined by structural and functional features. Finally, if the domains of superfamilies and families share same major secondary structural

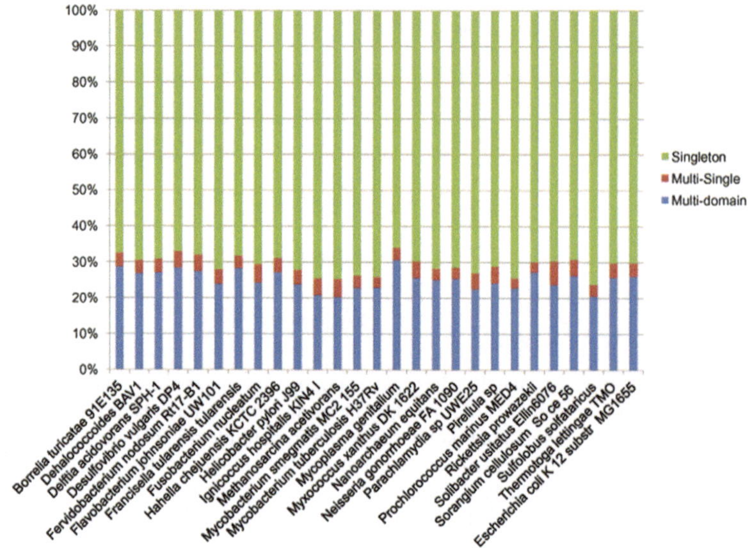

Fig. 2.5 A distribution of domains in various prokaryotic genomes. On an average 30 % of proteins possess multi-domain architecture. About 2–3 % of proteins possess repeats of the same domain (i.e. Multi-Single). Singletons are proteins with single domain. In order to identify domains, every protein sequence of organisms was searched against Pfam domain database using HMMER program [25–27]

elements in the same arrangement and with the same topological connections, they are classified as having a common "Fold" [23, 24].

A number of Web services have been developed for protein domain analysis and routinely used databases are Pfam, Structural Classification of Proteins (SCOP), ProDom, etc. [25, 28].

2.3 Multi-Facet Functional Aspects of Proteins

As mentioned earlier, protein function can be defined at many levels, such as protein domains, pathway, or subcellular localizations. We will study the distribution of some of the important functional aspects of proteins encoded by various prokaryotic organisms.

2.3.1 Subcellular Localization of Proteins

A significant portion of genes encodes for proteins that are essential as structural component and the transport-related functions of the cell. These proteins include

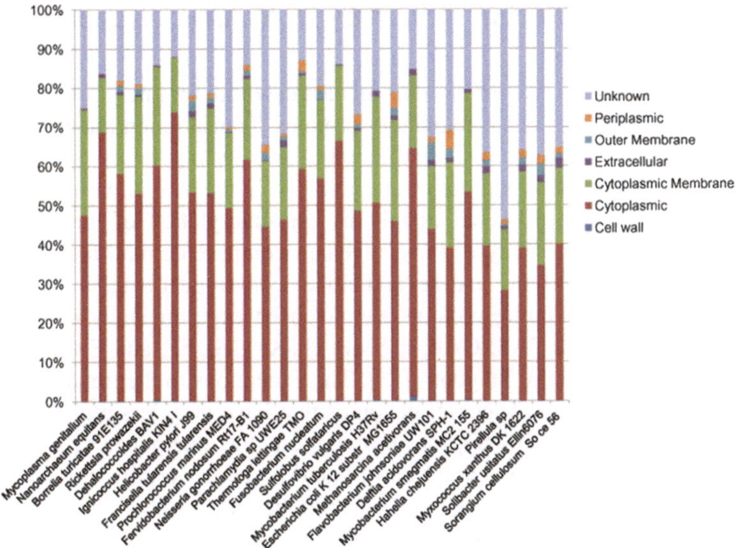

Fig. 2.6 Subcellular localization of proteins encoded by various prokaryotic genomes. The subcellular localization data for completely sequenced genomes was downloaded from PSORTb Web server. Unknown proteins that are not predicted with high confidence to one of the subcellular localization. Organisms are arranged according to their protein coding capacity. Fewer protein-encoding organisms start from left. On an average 56 % proteins are cytoplasmic followed by cytoplasmic membrane proteins. Many organisms do not encode for cell, outer membrane, and periplasmic proteins

transmembrane, periplasmic, outer membrane, extracellular proteins. These proteins show distinct sequence features than that of cytoplasmic proteins. These features such as charged amino acid, periodic stretches of hydrophobic amino acids in the protein sequences have been utilized to predict subcellular localization [29, 30]. Recently, a multi-subcellular localization method was also developed to chart all possible subcellular localization of genome-encoded proteins [30]. Figure 2.6 shows distribution of proteins that are known to be targeted to various subcellular locations in various prokaryotic organisms.

It can be observed from Fig. 2.6 around 56 % of proteins are predicted as localized in the cytoplasm, whereas 20–30 % are targeted to the cytoplasmic membrane. A small fraction of proteins seems to be targeted to the cell wall, outer membrane, and periplasmic space. Gram-positive bacteria such as Mycobacterium species lack these proteins. A significant portion of proteins could not be mapped to subcellular locations referred as "Unknown". Either these proteins are predicted with low confidences or they are assigned to multiple subcellular localization. Interestingly, their percentage goes higher as the number of proteins encoded by genomes goes up (Fig. 2.6). It can also be observed that the percentage of cytoplasmic proteins decreased with it. It is worth to explore whether it is an artifact of prediction algorithm or results are biologically significant.

2.3.2 Regulatory Protein Repertoire in Sequenced Genomes

One of the important classes of proteins that are essential for fine-grained control of overall activities of the cells is regulatory proteins. Transcription factors and signal transduction proteins are well studied among the other regulatory components of the cell. Signal transduction proteins are involved in the activation or inactivation of cascades of events triggered mainly by phosphorylation of their substrates. Comparative genomics analyses have revealed complex domain architectures of signal transduction proteins and several novel conserved domains. Most of the signaling domain containing proteins has not been experimentally characterized. Many domains such as PAS, GAF, HAMP, HisK, GGDEF, and EAL are widely spread across the bacterial kingdom [31]. The exceptions to the trend are some of the archaeal and parasitic bacterial genomes that have very less representation of signal transduction domains (Fig. 2.7). It suggests the emergence of signal transduction systems early in the evolution of bacteria. However, the presence of few instances in case of archaeal genomes could be the result of horizontal gene transfer [31]. Overall, the phylogenetic distribution of signaling domains is skewed. It has been observed that complexity of signaling systems differs even among closely related organisms. The complexity correlates with the organism's lifestyle, ecological niche, and typical environmental challenges it encounters [32]. Identification of signaling proteins becomes little bit tricky due to their modular or multi-domain nature since each domain evolves independently. In such cases homology-based tools such as BLAST has to be used cautiously. Signaling proteins are often analyzed using tools that predict domains such as HMMER [27]. Some of the good resources for signal protein analysis have been given in Table 2.1.

Another class of regulatory protein called as transcription factors (TFs) bind upstream region of their target genes in a sequence-dependent manner and regulate their expression. These factors can be activators or repressors of transcription, or both. These TFs can be grouped into three categories: (1) global transcription factors such as Crp, H–NS which regulates hundreds of genes, (2) local transcription factors which mostly regulate small sets of genes such as LacI; most often, they modulate expression of genes that are proximal to them on chromosome, (3) third category forms transcription regulators that connect global and local transcription factor activities and often referred to as intermediates. Due to such hierarchical activities of various TFs, the regulatory network forms three layers of hierarchical structure in which global regulators are at the top controlling major checkpoints and local regulators controlling local functional pathways. It has been observed that bacterial TFs have gone through flexible evolution. TFs evolve much faster than their target genes across three kingdoms of life [33, 34]. Surprisingly, global regulators are not conserved during evolution which may suggest a plasticity in rewiring of regulatory networks. In fact, comparative analysis has suggested that the transcriptional regulatory network is highly flexible as compared to the genetic components of organisms [33].

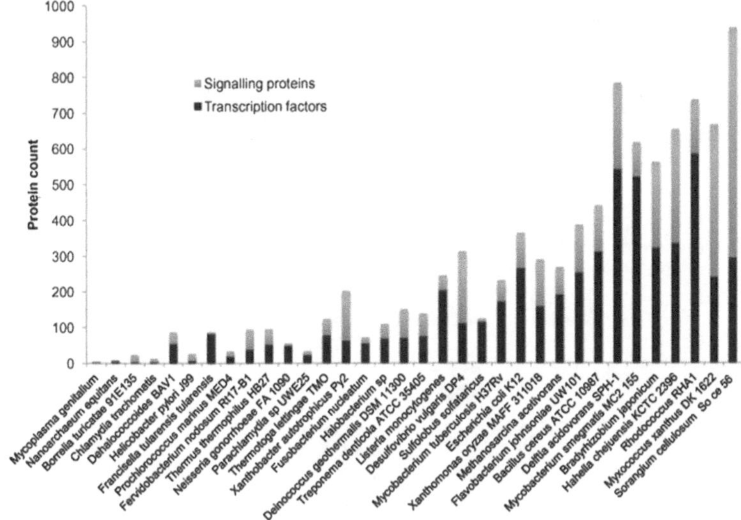

Fig. 2.7 A distribution of transcription factors and signaling proteins in various prokaryotic genomes. Organisms are arranged according to their protein coding capacity. Fewer protein-encoding organisms start from left. The figure reveals higher number of transcription factors and signaling proteins encoded from many genomes which encode higher number of proteins. However, this trend is not consistent

There is a close relationship between TFs and signaling proteins, since 66 % of response regulator output domains of signaling proteins seems to bind DNA. It suggests their ability to modulate transcriptional activity [35, 36]. As shown in Fig. 2.7, we do not find any correlation between the transcription factors and signaling proteins. Number of both classes differs dramatically in some of the genomes such as *Rhodococcus* and *Sorangium* species.

2.3.3 Metabolic Protein Repertoire

Comparative analysis of numerous protein sequences and structures combined with genome comparison has yielded new insights into the evolution of enzymes and their functions [37, 38]. In metabolic pathway proteins act together and thereby often show conserved phyletic pattern. Another evolutionary trend is observed from comparative analysis of enzymes which suggests the non-orthologous displacement of some enzymes by non-homologous proteins with the same function. These observations reflect divergence and convergence of enzyme function at different levels. Comparative microbial genomics suggests the correlation between genome size and metabolic diversity. Parasitic organisms encode fewer enzymes as compared to free-living organisms. Several databases report metabolic pathways of

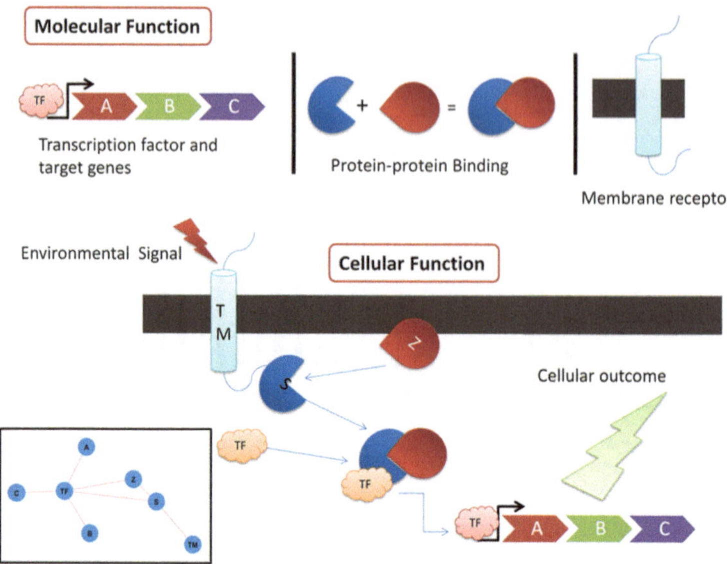

Fig. 2.8 A schematic representation of molecular and cellular function of proteins. A sum of parts reveals more information about functioning of biological process than the individual parts of cellular components. Upper panel of figures from left to right depict transcriptional regulator and its target genes, binding activity of two proteins, and a transmembrane receptor, respectively. Cellular function panel describes functional relationships of these proteins. Inset shows network/graph constructed based on proteins functional links

completely sequenced genomes such as KEGG, BioCyc, EcoCyc, etc. [39, 40] (Table 2.1). Many enzyme superfamilies tend to have maintained structure and sequence during evolution and phylogenetic analysis often leads us to define their function at substrate level.

2.4 Protein Function in Post-Genomic Era

As shown in Table 2.1, mostly, protein function prediction methods depend on the homologous relationship of unknown protein sequence with the experimentally characterized proteins. The data represented in Fig. 2.3 suggests around 40 % of proteins are still annotated as unknown or hypothetical even after application of cutting-edge homology-based methods. Moreover, homology-based methods cannot provide functional clues about the proteins that are not similar by sequence but acts in related biological processes/pathways. Therefore, the homology-based methods can only find molecular functions such as catalysis, binding to specific partner/ligand, etc., but little information about the cellular function or context of the protein.

The knowledge of protein's context is crucial since proteins virtually never work alone in cells, but often interact with many partners in order to perform the specific task. This connectivity among proteins that participate in a particular function is an important feature of cellular organization and regulation. Hence, a new class of computational methods has been developed that draw inferences about functional relationships between proteins by analyzing the context in which proteins are found [41–45]. Since these methods do not use homology to infer function, they are referred to as *nonhomology*-based protein function prediction methods [46]. These methods potentially predict cellular function of uncharacterized proteins but they can only provide pointers to the molecular function. Thereby, they can be applied for finding contextual protein functions, reconstructing cellular pathways, and revealing new metabolic pathways. Figure 2.8 shows how contextual information of protein helps to understand its system level role in a particular biological process as compared to its molecular function. The contextual information of every cellular protein can be conceptualized in the form of mathematical object known as *network or graph*, which is composed of nodes and edges (Fig. 2.8 inset). Nodes are the proteins and the edges joining them represent physical or functional interactions. The term 'network' has become popular in biological sciences well after the emergence of nonhomology-based methods for protein function predictions [47].

In the next chapter, we present a state of the art in the area of nonhomology-based protein function prediction methods: How the results of these methods led to define a global atlas of protein–protein interactions for an organism of interest and why the protein interaction predictions and analyses are the subject of widespread studies.

References

1. Rison, S.C., Hodgman, T.C., Thornton, J.M.: Comparison of functional annotation schemes for genomes. Funct. Integr. Genomics **1**(1), 56–69 (2000)
2. Whisstock, J.C., Lesk, A.M.: Prediction of protein function from protein sequence and structure. Q. Rev. Biophys. **36**(3), 307–340 (2003)
3. Janga, S.C., J.J. Diaz-Mejia, Moreno-Hagelsieb, G.: Network-based function prediction and interactomics: the case for metabolic enzymes. Metab Eng. **13**(1), 1–10 (2011)
4. Camon, E., et al.: The gene ontology annotation (GOA) database: sharing knowledge in uniprot with gene ontology. Nucleic Acids. Res. **32**(Database issue), D262–266 (2004)
5. Harris, M.A., et al.: The gene ontology (GO) database and informatics resource. Nucleic Acids Res. **32**(Database issue), D258-D261 (2004)
6. Bernal, A., Ear, U., Kyrpides, N.: Genomes online database (GOLD): a monitor of genome projects world-wide. Nucleic Acids Res. **29**(1), 126–127 (2001)
7. Schneiker, S., et al.: Complete genome sequence of the myxobacterium *Sorangium cellulosum*. Nat. Biotechnol. **25**(11), 1281–1289 (2007)
8. Lopez-Madrigal, S., et al.: *Complete genome sequence of "Candidatus Tremblaya princeps" strain PCVAL, an intriguing translational machine below the living-cell status.* J Bacteriol. **193**(19): p. 5587-8
9. Pearson, W.R.: Comparison of methods for searching protein sequence databases. Protein Sci. **4**(6), 1145–1160 (1995)

10. Altschul, S.F., et al.: Gapped BLAST and PSI-BLAST: a new generation of protein database search programs. Nucleic Acids Res. **25**(17), 3389–3402 (1997)
11. Adams, M.A., et al.: Piecing together the structure-function puzzle: experiences in structure-based functional annotation of hypothetical proteins. Proteomics **7**(16), 2920–2932 (2007)
12. Gotoh, O.: Multiple sequence alignment: algorithms and applications. Adv. Biophys. **36**, 159–206 (1999)
13. Procter, J.B., et al.: Visualization of multiple alignments, phylogenies and gene family evolution. Nat. Methods **7**(3 Suppl), S16–S25
14. Koonin, E.V.: Orthologs, paralogs, and evolutionary genomics. Annu. Rev. Genet. **39**, 309–338 (2005)
15. Pazos, F., Valencia, A.: Protein co-evolution, co-adaptation and interactions. EMBO J. **27**(20), 2648–2655 (2008)
16. Koonin, E.V.: Obituary: Walter Fitch and the orthology paradigm. Brief Bioinform. **12**(5), 377–378 (2011)
17. Thompson, J.D., Gibson T.J., Higgins D.G., Multiple sequence alignment using ClustalW and ClustalX. Curr. Protoc. Bioinform. **Chapter 2**(Unit 2), 3 (2002)
18. Apic, G., Gough, J., Teichmann, S.A.: Domain combinations in archaeal, eubacterial and eukaryotic proteomes. J. Mol. Biol. **310**(2), 311–325 (2001)
19. Marsh, J.A., Teichmann, S.A.: How do proteins gain new domains? Genome Biol. **11**(7), 126
20. Vogel, C., et al.: Structure, function and evolution of multidomain proteins. Curr. Opin. Struct. Biol. **14**(2), 208–216 (2004)
21. Vogel, C., Teichmann, S.A., Pereira-Leal, J.: The relationship between domain duplication and recombination. J. Mol. Biol. **346**(1), 355–365 (2005)
22. Chothia, C., et al.: Evolution of the protein repertoire. Science **300**(5626), 1701–1703 (2003)
23. Cosentino Lagomarsino, M., et al.: Universal features in the genome-level evolution of protein domains. Genome Biol. **10**(1), R12 (2009)
24. Apic, G., Gough, J., Teichmann, S.A.: An insight into domain combinations. Bioinformatics **17**(Suppl 1), S83–S89 (2001)
25. Bateman, A., et al.: The Pfam protein families database. Nucleic Acids Res. **32**(Database issue), D138-D141 (2004)
26. Sonnhammer, E.L., Eddy, S.R., Durbin, R.: Pfam: a comprehensive database of protein domain families based on seed alignments. Proteins **28**(3), 405–420 (1997)
27. Finn, R.D., Clements J., Eddy, S.R.: HMMER web server: interactive sequence similarity searching. Nucleic Acids Res. **39**(Web Server issue), W29–W37 (2011)
28. Andreeva, A., et al.: Data growth and its impact on the SCOP database: new developments. Nucleic Acids Res. **36**(Database issue), D419–D425 (2008)
29. Kall, L., Krogh, A., Sonnhammer, E.L.: Advantages of combined transmembrane topology and signal peptide prediction—the Phobius web server. Nucleic Acids Res. **35**(Web Server issue), W429–W432 (2007)
30. Yu, N.Y., et al.: PSORTb 3.0: improved protein subcellular localization prediction with refined localization subcategories and predictive capabilities for all prokaryotes. Bioinformatics **26**(13), 1608–1615 (2010)
31. Galperin, M.Y., Nikolskaya, A.N., Koonin, E.V.: Novel domains of the prokaryotic two-component signal transduction systems. FEMS Microbiol. Lett. **203**(1), 11–21 (2001)
32. Galperin, M.Y.: A census of membrane-bound and intracellular signal transduction proteins in bacteria: bacterial IQ, extroverts and introverts. BMC Microbiol. **5**, 35 (2005)
33. Lozada-Chavez, I., Janga, S.C., Collado-Vides, J.: Bacterial regulatory networks are extremely flexible in evolution. Nucleic Acids Res. **34**(12), 3434–3445 (2006)
34. Kummerfeld, S.K., Teichmann, S.A.: DBD: a transcription factor prediction database. Nucleic Acids Res. **34**(Database issue), D74–D81 (2006)
35. Galperin, M.Y.: Structural classification of bacterial response regulators: diversity of output domains and domain combinations. J. Bacteriol. **188**(12), 4169–4182 (2006)
36. Bourret, R.B.: Census of prokaryotic senses. J. Bacteriol. **188**(12), 4165–4168 (2006)

37. Galperin, M.Y., Koonin, E.V.: Functional genomics and enzyme evolution. Homologous and analogous enzymes encoded in microbial genomes. Genetica **106**(1–2), 159–170 (1999)
38. Galperin, M.Y., Koonin, E.V.: Divergence and convergence in enzyme evolution. J. Biol. Chem. **287**(1), 21–28 (2012)
39. Kanehisa, M., et al.: The KEGG resource for deciphering the genome. Nucleic Acids Res. **32**(Database issue), D277–D280 (2004)
40. Keseler, I.M., et al.: EcoCyc: a comprehensive view of *Escherichia coli* biology. Nucleic Acids Res. **37**(Database issue), D464–D470 (2009)
41. Gaasterland, T., Ragan, M.A.: Microbial genescapes: phyletic and functional patterns of ORF distribution among prokaryotes. Microb. Comp. Genomics **3**(4), 199–217 (1998)
42. Galperin, M.Y., Koonin, E.V.: Who's your neighbor? New computational approaches for functional genomics. Nat. Biotechnol. **18**(6), 609–613 (2000)
43. Dandekar, T., et al.: Conservation of gene order: a fingerprint of proteins that physically interact. Trends Biochem. Sci. **23**(9), 324–328 (1998)
44. Galperin, M.Y., Walker, D.R., Koonin, E.V.: Analogous enzymes: independent inventions in enzyme evolution. Genome Res. **8**(8), 779–790 (1998)
45. Tamames, J., et al.: Conserved clusters of functionally related genes in two bacterial genomes. J. Mol. Evol. **44**(1), 66–73 (1997)
46. Marcotte, E.M.: Computational genetics: finding protein function by nonhomology methods. Curr. Opin. Struct. Biol. **10**(3), 359–365 (2000)
47. Barabasi, A.L., Oltvai, Z.N.: Network biology: understanding the cell's functional organization. Nat. Rev. Genet. **5**(2), 101–113 (2004)

Chapter 3
Co-Evolutionary Signals Within Genome Sequences Reflect Functional Dependence of Proteins

Abstract In the course of evolution, proteins involved in a particular biological process or pathway are often subjected to the same selection pressure and adaptive constraints through various molecular mechanisms. Thus, proteins that are working together in the cell often co-evolve and show similar evolutionary trajectories. One of the evolutionary constraints that act on functionally related proteins, is the concerted appearance of genes encoding them in the organisms for which their function is indispensable, and disappearance otherwise. Likewise, physically interacting proteins are expected to have correlated mutations in their sequences and/or nucleotide sequence of genes encoding them in order to maintain binding interfaces. These two forms of co-evolutionary behavior of genes and their products in order to maintain their function leave pattern over the long evolutionary periods. In the post-genomic era, these co-evolutionary patterns have been utilized to reconstruct genome-scale protein–protein interactions and biological pathways using various methods. In this review, we have described the basic principles of these methods and the novel strategies to improve their prediction qualities.

3.1 Introduction

The species thriving in the specific ecological niche evolve phenotypes or traits, which can help them withstand the surrounding environmental conditions. Consequently, genes and their products responsible for expression of these traits are expected to show cooperative evolution due to the selection imposed by surrounding conditions. This assumption can be explained by the fact that prokaryotic genomes are often subjected to loss or gain of genes through various molecular mechanisms during evolution [1–6]. The major forces significantly contributing to the microbial genome evolution are the acquisition of genes through horizontal gene transfer and

V. Y. Muley and V. Acharya, *Genome-Wide Prediction and Analysis of Protein–Protein Functional Linkages in Bacteria*, SpringerBriefs in Systems Biology, DOI: 10.1007/978-1-4614-4705-4_3,

the loss of genes through reductive evolutionary processes. As population diverges, the acquired genes would likely be fixed in the population if they are advantageous. Otherwise, acquired genes without any selection pressure are likely to be lost over short evolutionary periods or converge to some other functions. In other words, the evolution of genes and their products is constrained by their functions. As a result, genes involved in a particular biological process or pathway often show similar evolutionary trajectories and hence indicates their co-evolution. This basic premise has significantly contributed to our basic understanding of evolutionary and functional constraints acting on the genes.

With the availability of several completely sequenced genomes, we are now in the position to quantify the co-evolutionary behavior of genes at genome-scale. Eventually, co-evolution of the genes and their products often indicates functional and/or physical interactions among them, thereby revealing higher level organization of biological systems. The past decade has witnessed significant progress in predicting functional and physical protein–protein interactions (PPIs) through methods that have been developed for measuring co-evolutionary signals in the genomic sequences. Let us discuss some of these methods and the basic principles behind their success in quantifying co-evolution. For brevity, both the functional and physical interactions between proteins hereafter are referred as PPIs unless mentioned explicitly.

3.2 Co-Occurrence of Proteins as an Indicator of Functional Linkage

One of the simplest yet powerful methods to predict functionally linked proteins (or genes) is the co-occurrence of two proteins across multiple genomes called phylogenetic profiling [1, 7]. During the reductive evolutionary process, if one of the two interdependent proteins is lost for any reason, the evolutionary pressure to maintain the other is no longer needed as it cannot work alone. Eventually, in subsequent generation, it may get lost unless converged to some other functional role. Likewise, one of the two interdependent proteins is ‘acquired’ (i.e. horizontal gene transfer) during evolution; the other partner has to be acquired in order to maintain the functions mutually driven by them. In practical terms, all this means that proteins involved in a particular function will tend to be present in the subset of genomes where that function is essential and hence co-evolving under the same functional constraints while absent in the rest [1, 6, 8, 9]. For instance, several bacteria exhibit motility phenotype whereas others do not, comparative genomics of these organisms revealed around 50 proteins that are exclusively present in motile but not in the non-motile organisms [10]. The majority of these proteins are involved either in the chemotaxis or in the biosynthesis of flagella apparatus, and both processes are indispensable for motility. Many Gram-positive bacteria have inherited a set of proteins involved in endospore formation. Thus, these proteins co-occur only in the spore forming organisms and are totally absent from the non-sporulating organisms [10, 11].

3.2.1 Detection of Orthologs

In order to detect proteins encoded from *query genome* (i.e. genome of interest) that are co-occurring, we require their orthologs from various other genomes. These other genomes are commonly referred to as *reference* or target. Briefly, orthologs are homologous proteins that are a result of speciation event and perform analogous functions in different organisms. The other form of homologous proteins is paralogs that are a result of duplication event and they may perform similar functions in the same organism. The detection of homologous proteins as either paralogs or orthologs is a difficult problem in the absence of correct speciation history of organisms [12]. This problem can be overcome to some extent by identifying 'bidirectional best hit', i.e., given two proteins $Q1$ and $R2$ from two genomes Q and R, we refer to them as orthologs if $Q1$ is the best match for $R2$ when searched against genome R, and $R2$ is the best match for $Q1$ when searched against genome Q [1, 13, 14]. Therefore, the presence of ortholog for a query protein ensures the possibility of its function encoded from the corresponding reference genome.

3.2.2 Computing Phylogenetic Profiles

The presence and absence of orthologs of the query proteins across reference genomes is represented as vectors called phylogenetic profiles (PPs) or phyletic patterns or co-occurrence profiles, and an approach is referred to as phylogenetic profiling [1, 15]. A schematic representation of PP is given in Fig. 3.1. Originally, the PP of a protein was represented qualitatively as a binary vector, where '1' represented the presence of the protein in a reference genome and '0' represented its absence [15]. Similarly, the presence of a given protein in the PP can also be quantitatively represented in the form of transformed e-value scores of sequence alignments with their orthologs [16]. Moreover, the presence of a given protein in the PP can be represented in the form of a bit score [17]. The approach is relatively less popular as compared to the e-value-based or the binary representation of profiles. However, the bit score is a normalized alignment score of a given query protein and its homolog in the reference genome, which reflects the extent of sequence similarity between them [18]. The profiles constructed using either e-value or bit score are referred as similarity profiles rather than co-occurrence since they represent sequence divergence information quantitatively [7]. Eventually, the representation of the presence and absence of orthologs for all proteins of query genome leads to the formation of matrix. Rows in the matrix are proteins of query genome, 1, 2, 3, …, i and columns are reference genomes, 1, 2, 3 …, j, where i is the number of proteins in a query genome and j is the number of reference genomes.

		G1	G2	G3	G4	G5	G6	G7	G8	G9	G10	G11	G12	G13	G14	G15	G16
Arginine Metabolism	argA						13				72	55					
	argB		18	24	21	27		27	25	24	73	23	26				
	argC	22	31	39	32	39	40	34	35	30	72	33	24				
	argD	38		40		42	55	41			72	72					
	argE			15		16					75	50					
Flagella	flgK					12	9	10	14	13	62	24					
	flgL					17		19	27	12	62	14					
	flhA					45	45	29	43	26	73	52					
	flhB					43	36	29	35	31	73	38					
	fliC					17	17	15	19	18	38	18					
	fliD							15	11	10	37	18					
	fliF					18	23	11	17	24	67	33					
	fliG					35	41	23	39	39	73	48					
	fliI					56	58	37	56	48	72	63					
	fliM					25	20	10	25	24	75	40					
	fliP					61	59	43	60	64	72	62					
Lipid A	lpxA									50	74	56	43	71	49		
	lpxB									18	74	52	22	63	41		
	lpxC									49	76	62	27	78	50		
	lpxD								11	34	76	54	31	67	51		
DNA Repair	uvrA	77	72	74	72	76	72	72	74	58	75	84	66	83	72	65	66
	uvrB	70	73	74	71	75	74	71	74	65	73	76	69	78	70	60	59
	uvrC	21	37	36	27	38	37	36	38	24	73	54	32	66	45	31	26
	uvrD	44	42	46	43	47	41	45	46		75	67	35	67	47	29	32
						Motile Organisms											

Fig. 3.1 A schematic representation phylogenetic profiling method for predicting protein–protein interactions. Phylogenetic profiling assumes co-occurrence of proteins across various genomes. For example, Flagella proteins show strong co-occurrence pattern across motile organisms. The matrix represents phylogenetic profiles of *Escherichia coli* proteins belonging to four functional categories. Each *row* is protein and *column* represents genome. Columns *G*1–*G*8 (*red color*) represent Gram-positive genomes while *G*9–*G*16 are Gram-negative genomes. The presence of *Escherichia coli* protein in a genome is represented with bit score of alignment with corresponding ortholog, otherwise element of matrix is blank

Each (i, j) cell of this matrix can be filled with the '1' or bit score or *e*-value of query genome protein *i* and its ortholog in the *j*th reference genome. If a protein is absent in any reference genome then it is denoted with score zero (Fig. 3.1).

3.2.3 Computing Co-Evolution

Once PP matrix is created, then the similarity of each protein profile with the remaining protein profiles can be computed. The similarity score between PPs of two proteins represents the extent of their co-evolutionary behavior. Several metrics have been proposed to compute similarity between two profiles. However, all these metrics rank the protein pairs in the same order [7]. The most popular metrics are mutual information (MI) and Pearson correlation coefficient (PCC) which are described below.

3.2.3.1 Mutual Information

MI measures the information content that PP of two proteins conveys about each other. MI for binary PP of two proteins, *X* and *Y* can be calculated as follows [16]:

$$\mathrm{MI}(X, Y) = H(X) + H(Y) - H(X, Y)$$

where, $H(X)$ and $H(Y)$ are the *empirical* information entropy of PP of X and Y proteins. $H(X, Y)$ is the joint information entropy of PP of X and Y proteins, respectively. These expressions are defined as

$$H(X) = \sum_{x} \frac{x}{N}$$

where, 'x' is the frequency with which x's is observed in PP of X. For binary profile these values of x would be only two, i.e., 1 and 0. N is the number of reference genomes.

$$H(X, Y) = -\sum_{x,y} \frac{(x, y)}{N} \log \frac{(x, y)}{N}$$

where, (x, y) is the frequency with which the pair of values (x, y) are observed in PP of X and Y in the vector position of corresponding reference genomes. For binary profile values of (x, y) would be 4, i.e., ('1','1') if both proteins are present in the Nth reference genome; ('1','0'), if X is present and Y is absent; ('0','1'), if X is absent and Y is present; ('0','0'), if both proteins are absent from Nth reference genome. In case one has constructed PP entries with real-value numbers such as *e*-value or bit score values then the best way is to represent them by binning in intervals of 0.1. The joint entropy is minimal at zero when X and Y consist either entirely of zeros or entirely of ones. In contrast, $H(X, Y)$ is maximal if (X, Y) has an equal number of zeros and ones.

3.2.3.2 Pearson Correlation Coefficient

PCC measures the degree of linearity between two profiles [7, 17] as follows:

$$r(X, Y) = \frac{\sum (X_i - \overline{X})(Y_i - \overline{Y})}{\sqrt{\sum (X_i - \overline{X})^2 (Y_i - \overline{Y})^2}}$$

where, 'i' is a length of PP, i.e, the number of reference genomes. PCC measures similarity between two profiles on a scale between −1 and 1. The scores toward −1 reflect anti-correlated proteins and toward one reflect co-evolution. The scores around 0 reflect that the occurrence of two proteins in reference genomes is random.

The biggest advantage of PCC is its ability to capture anti-correlated protein pairs. It means whenever protein X is absent from any reference genome then Y is present and vice versa. Therefore, it indicates that the role of the protein that is absent from reference genome is likely to be fulfilled by its anti-correlated partner. Many previous analyses have shown that such protein pairs perform similar functions, and this phenomenon is referred to as a *non-orthologous displacement* [9, 19, 20].

Despite the successful application of these metrics in predicting a functional relationship between proteins, they violate the evolutionary relationship among species by assuming that the occurrence of an ortholog in one species is independent of its occurrence in another [7]. This violation to some extent can be overcome by using explicit models of evolution to infer gain and loss events of two proteins on branches of species tree [21]. It reduces the high rates of false predictions observed in conventional approaches. However, these methods reconstruct ancestral states using concurrence patterns and assume they are not erroneous; this may not always be the case [7, 21]. Moreover, these approaches are computationally expensive.

3.2.4 Scope and Future Perspectives

Despite the simplicity of PP and the most widely used method, there have been controversial observations with respect to the reference genome selection [22–25]. Selection of reference genome for profiling query genome proteins is one of the critical steps, which determines the accuracy of predicted functional linkages. For instance, PP constructed using reference set of phylogenetically close relatives of query genome may falsely predict interactions. Query genome would likely share many common lineage-specific proteins due to their close relatedness and hence result into similar PPs for a number of proteins irrespective of their functional relevance [40, 57]. Most of the previous studies have agreed upon phylogenetically diverse sets of reference genomes for PP to achieve high-quality predictions. In our opinion, the consideration of the phylogenetically distant relatives of query genome as reference is leading us to infer functional relationships among the proteins that are dominant in the three domains of life. In other words, we have restricted our search space to the proteins that are likely to be present in the majority of organisms. Thus, we are missing the functional relationship between many specialized processes that are lineage or ecological niche specific. Moreover, the linkages of these specialized processes to that of housekeeping or processes that are dominant in major living forms. The prediction of the interactions among the lineage or ecological niche specific proteins and with other housekeeping proteins promises to shed light on the organization and regulation of the specialized processes. Unfortunately, our inability to remove functional information from the speciation or evolutionary relationship between reference genomes is quite limited. Many pathogenic species have evolved with specialized systems to infect their host. We have very limited knowledge about these systems despite the sequencing of several sub-species, strains, or serotype for many important pathogens. For example, the complete genome sequences are available for 29 pathogenic and non-pathogenic species of Mycobacterium. These pathogenic species are responsible for important diseases like tuberculosis (TB), leprosy, and skin lesions. However, we know little about the specialized processes that occur in these deadly pathogens. Therefore, the development of methods, which can remove speciation information, can make great impact in understanding the organization of specialized processes.

In this direction, it is unexplainable why previous analyses have preferred *e*-values or binary digits to construct profiles when bit score profiles outperformed them [17, 26]. Enault et al. considered double normalization of bit scores for construction of PP [17]. The gain of 25 % increased enzymes identification and reduction of false predictions with 20 % margin was observed as compared to the profiles constructed using binary digits. Furthermore, previous studies have used the transformed *e*-values to create PPs since the authors believed that *e*-value measures sequence divergence [16, 23, 24, 27]. We do not agree with the notion that *e*-value measures the sequence divergence information. Because, by definition *e*-value is a measure of probability that a given Basic Local Alignment Search Tool (BLAST) search hit is obtained by chance for a size of given database. Bit score is a normalized sequence score representing the quality of match based on sequence alignment of query protein and its ortholog [18]. Thus, bit scores truly reflect the sequence divergence information and not the *e*-value. Therefore, in our opinion, it is preferable to construct PP using bit scores as opposed to transformed *e*-values to capture sequence divergence in a better way [52]. Our recent study suggests that bit score based PPs minimize the effect of reference genome selection. We have demonstrated the effectiveness of normalized PPs constructed using bit scores in predicting PPIs using closely related reference genomes [52]. We believe our results indicate the possibility of inferring PPIs involved in specialized biological systems.

3.3 Similarity of Phylogenetic Trees of Proteins as an Indicator of Functional Linkage

We have mentioned above that the genes and their products show cooperative evolution due to the selection imposed by the surrounding conditions. Therefore, the functional constraints on the set of genes drives their co-evolution. It is also discussed how one can quantify co-evolution of proteins using co-occurrence analysis. Pazos and Valencia have extrapolated the concept of co-evolution at the molecular level to pioneer a class of computational methods called mirrortree that predicts physical PPIs [28]. It has to be noted that phylogenetic profiling cannot distinguish between physical and functional PPI but mirrortree-based methods do. The mirrortree-based methods assume resemblance between phylogenetic trees of physically interacting protein families. In order to understand the concept of co-evolution for predicting PPIs using mirrortree method, we have to begin with the composition of amino acids in the protein sequence.

3.3.1 Computing Co-Evolution at the Residue Level

The co-evolution is strictly defined as the cooperative evolution of a species in response to selection imposed by another due to ecological constraints. This definition is applicable for protein sequences too, for example, when two amino acid residues say '*P*' and '*Q*' are making contact in the protein spatial structure. An unfavorable amino acid change at site '*P*' may go without negative consequences, if '*Q*' is simultaneously mutated in such a way that the original contact is rescued. Such mutations have been described as compensatory mutations or correlated mutations and amino acid residues were referred as co-evolving. The starting point of all methods that determine co-evolving residues in the protein sequence is a multiple sequence alignment (MSA). In the MSA, homologous protein sequences of N number of organisms can be arranged in a matrix of *N* rows and *L* columns, in such a way that equivalent residues are placed in the same columns, to best represent the evolutionary relationships among the sequences. Hence, a column in an MSA represents amino acid changes accepted during evolution at that position. A co-variation of 20 amino acid residues at various positions in an MSA reflects their co-evolutionary behaviors. Since the development of the first approach to detect co-evolving residues in an MSA in 1994 by Valencia et al. [29], several methods have been proposed to evaluate their significance in predicting residue–residue contacts for ab initio structure prediction. However, this application is strongly limited by the fact that accuracies of these methods to predict structural contacts hardly exceed 20 % [30]. Impressive studies by Ranganathan et al. suggested an unexpected degree of simplicity in amino acid interactions in the atomic structure of protein. They observed many direct packing interactions between residues that are not co-evolving and some distant sites linked through networks of co-evolving residues are predicted to be coupled using statistical coupling analysis [30, 31]. In addition, the co-evolving residues in the protein sequence were often observed nearby functionally important sites such as active or ligand binding sites [32, 33], in fact certain co-evolving residues are more likely to be ligand-binding or functional sites [34].

The relationship between correlated mutations and functional sites has not only been found for the intra-protein residues but also between residues in different proteins [28, 33, 35, 36]. Correlated mutations among proteins that form obligate complexes have been observed to be more evident [37], even though the corresponding residues often do not form direct physical contacts [33, 38].

3.3.2 Computing Co-Evolution at the Protein Level

According to Valencia et al. [39], the similarity of the phylogenetic trees of interacting protein families is the most closely related protein feature that follows the original definition of co-evolution. For instance, phylogenetic trees of ParC and ParE protein families have been shown in Fig. 3.2. These two proteins form active topoisomerase IV complex and belong to the type II topoisomerase family. Topoisomerase IV plays an essential role by removing double-stranded DNA crossings while progression of the replication fork and the chromosome segregation after replication [40, 41]. It decatenates the two daughter molecules after DNA replication. Moreover, topoisomerase IV complex is able to relax positive-DNA supercoils 20-fold faster than negative supercoils [42]. The phylogenetic tree

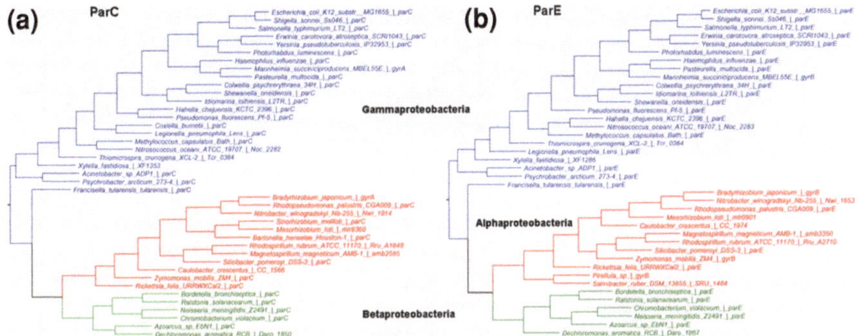

Fig. 3.2 A comparison of phylogenetic trees of protein ParC and ParE. Phylogenetic trees of ParC and ParE proteins show almost similar topology. These proteins interact physically with each other. The orthologs of *Escherichia coli* proteins ParC and ParE identified using best bidirectional hit approach in 121 reference genomes and aligned using ClustalW. Multiple sequence alignment then used to reconstruct neighbor-joining tree

topology of ParC and ParE protein families is almost identical. In the past, similarities between phylogenetic trees have also been observed for many receptors and their ligands, e.g., insulin, vasopressin, and their receptors [43, 44].

A topological similarity between phylogenetic trees of two protein families is possible only when, the similar amino acid substitutions have occurred during evolution in both proteins. Hence, one can assume the possibility of many positions of correlated amino acids substitutions in their MSAs and that could be the reason for their strong co-evolutionary behavior. Initial thoughts for similar phylogenetic trees were along the lines of co-evolution/correlated mutations of the interface or binding site residues, and the functional dependence of two proteins since these proteins interact physically [37, 45, 46]. The similarity of phylogenetic trees of interacting proteins has been quantified using available PPI datasets [46, 47]. This led to the development of the *mirrortree* method for predicting physically interacting proteins [47]. It is very difficult if not impossible to compare the topology of phylogenetic trees constructed for biological sequences and this problem is not fully solved. The MSA of the protein sequences provides enough information to capture the similarity of phylogenetic trees and hence their co-evolution. A schematic representation of the overall approach is given in Fig. 3.3 and the method is briefly discussed subsequently.

3.3.2.1 Mirrortree Method

Let us assume we expect two proteins *X* and *Y* to physically interact with each other. We can quantify their co-evolution at the level of amino acid sequences in three steps.

- Construct MSAs for protein *X* and *Y* using their orthologs from reference genomes. In order to get better results, it is better to choose proteins with

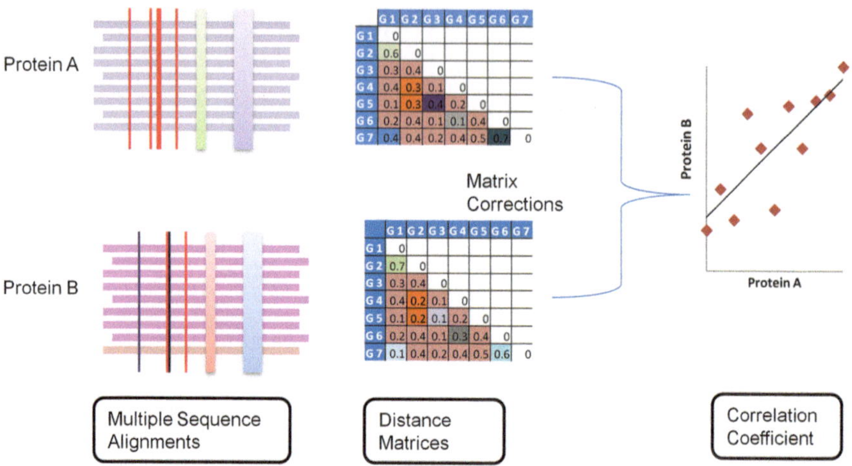

Fig. 3.3 A schematic representation of the mirrortree approach. Mirrortree compares distance matrices derived from aligned orthologs of query proteins. Prior to comparison, one can correct these matrices to exclude speciation information using the new approach. Then, correlation coefficient can be calculated reflecting the strength of co-evolution of protein *A* and *B*

orthologs from more than 15 genomes for MSA construction. The same criterion has been used in most of the previous studies.

- Then compute distance matrices for these two proteins using their MSA
  - Each protein matrix would be of size $n \times n$, where n represents the number of reference genomes in which orthologs are detected.
  - An element of the distance matrix, let us say $DX(i, j)$, represents the genetic distance between reference genomes i and j, which is a difference in amino acid sequences of protein '*X*' from reference genome i and j.
  - Distance matrices of two proteins *X* and *Y* are only comparable when their dimensions are the same. However, the dimension of each protein matrix may differ depending on its phyletic distribution in reference genome set.
  - Therefore, we keep only distances between reference genomes in which orthologs of both proteins are identified.
  - A minimum of 15 common reference genomes between distance matrices of both proteins should be good enough to capture their co-evolutionary behavior.
- The next step is to simply compute PCC between distance matrices of two proteins.

The degree of correlation between the distance matrices of two proteins quantifies the strength of co-evolving residues in them, and hence indirectly quantifies the similarity of phylogenetic tree. There are many programs available for MSA construction but we prefer to use ClustalW, since it has in-built application for distance matrix creation [48]. Instead of MSA, it is also possible to derive distance matrices using actual phylogenetic trees of proteins constructed using neighbor-joining method available in ClustalW program [48].

3.3.3 Scope and Future Perspective

To the best of our knowledge, mirrortree-based methods are ones that reliably predict physical PPI. Thereby, these methods have the capacity to infer interactions between subunits of protein complexes. However, one of the limitations of mirrortree methods is that the similarity of phylogenetic trees of the two proteins is manifestation of underlying speciation events [45, 49]. Therefore, the parameters where similarities of the phylogenetic trees of two proteins are also considered viz., the signals, about the evolutionary relationships among the common organisms used for analysis, as well as information about their physical interaction. This kind of global evolutionary relatedness of the organisms in the phylogenetic trees of proteins is called "background similarity" [45, 49]. Thereby, there is more chance that mirror tree method predicts a large number of false PPIs due to background similarity. In order to remove this background noise from distance matrices of proteins, researchers have further used molecular markers of tree of life such as 16S rRNA [49], or it can be nullified by normalization of distance matrices [45, 50]. The results obtained from these studies indeed showed improvement in the prediction accuracy over the original approach and significantly reduced falsely predicted interactions [45, 49, 50].

In our opinion, exclusion of background similarity using 16S rRNA as a marker of global evolutionary is quite easy and computationally less expensive. As explained above, the idea here is to derive distance matrix for 16S rRNA sequences from the reference genomes that have been used for ortholog detection. Then, the 16S rRNA distance matrix values between reference genomes need to be subtracted from the corresponding protein distance matrices. Thereby, the resulting protein distance matrices are expected to have information only due to functional relatedness than the background similarity. The reason behind this expectation is that 16S rRNA genes are one of the best conserved elements in prokaryotic organisms which is used for species identification. As a result, 16S rRNA distances between the closely related species would be far higher than the distantly related ones. Therefore, the subtraction of these distances from the protein matrices would reduce corresponding values significantly for closely related species as compared to distant ones. Although the scales of distances for proteins and 16S rRNA sequences are different and should be scaled before corrections/subtractions. Recently, the multidimensional scaling and superimposition approaches have been used to measure the global similarity between trees as well as incongruities between them [50]. Although, most of these methods have been shown to perform better than the original mirrortree method, the improvement is still marginal.

Therefore, the best use of this method in our opinion will come after solving the problem of background similarity due to global relatedness of species under consideration. In the case of phylogenetic profiling method, we have already mentioned the bias caused by inclusion of closely related species. The majority of researchers have agreed upon the utility of phylogenetically diverse reference genomes to overcome this problem. The fact is that most of these studies have been evaluated with gold standards, which are already biased toward well-studied proteins. These proteins are often present majority of several known genomes. Therefore, there is a margin for belief that the actual contribution of closely related genomes in predicting PPI is shaded due to gold standard datasets consisting of interactions among proteins from phylogenetically diverse genomes.

Despite limitations, mirrortree-based methods are the most reliable among others for predicting physical PPI. Thereby, these methods are promising candidates to modify them for predicting functional and ligand-binding sites. This field is still unexplored and is in its infancy. The power of the mirrortree in combination with other methods such as phylogenetic profiling can solve the problem of false predictions by these methods alone.

Finally, as mentioned above, one of the assumptions behind the ability of mirrortree-based methods to predict interaction between the two protein families was initially thought to be due to the correlated mutations at their binding interfaces, which was questioned in some of the recent studies [38, 51]. Further evidence has come from the study by Kann et al. which shows that binding interfaces along with neighborhood residues have higher co-evolutionary signal compared to that of regions outside binding interfaces [33]. They further demonstrated that some co-evolutionary signal remains in the protein sequences even after removal of binding neighborhood. It led to the conclusion that the correlation of phylogenetic trees of interacting proteins is not only due to the correlated mutations at their binding interfaces but is also contributed by common evolutionary pressure exerted on the whole protein sequence [33].

References

1. Tatusov, R.L., Koonin, E.V., Lipman, D.J.: A genomic perspective on protein families. Science **278**(5338), 631–637 (1997)
2. Aravind, L., et al.: Evidence for massive gene exchange between archaeal and bacterial hyperthermophiles. Trends Genet. **14**(11), 442–444 (1998)
3. Watanabe, H., et al.: Genome plasticity as a paradigm of eubacteria evolution. J. Mol. Evol. **44**(Suppl 1), S57–S64 (1997)
4. Koonin, E.V.: Evolution of genome architecture. Int. J. Biochem. Cell Biol. **41**(2), 298–306 (2009)
5. Lawrence, J.G.: Selfish operons and speciation by gene transfer. Trends Microbiol. **5**(9), 355–359 (1997)
6. Gaasterland, T., Ragan, M.A.: Microbial genescapes: phyletic and functional patterns of ORF distribution among prokaryotes. Microb. Comp. Genomics **3**(4), 199–217 (1998)
7. Kensche, P.R., et al.: Practical and theoretical advances in predicting the function of a protein by its phylogenetic distribution. J. R. Soc. Interface **5**(19), 151–170 (2008)

8. Koonin, E.V., Mushegian, A.R.: Complete genome sequences of cellular life forms: glimpses of theoretical evolutionary genomics. Curr. Opin. Genet. Dev. **6**(6), 757–762 (1996)
9. Koonin, E.V., Mushegian, A.R., Bork, P.: Non-orthologous gene displacement. Trends Genet. **12**(9), 334–336 (1996)
10. Slonim, N., Elemento, O., Tavazoie, S.: Ab initio genotype-phenotype association reveals intrinsic modularity in genetic networks. Mol. Syst. Biol. **2**, 2006 0005 (2006)
11. Singh, A.H., et al.: Modularity of stress response evolution. Proc. Natl. Acad. Sci. USA **105**(21), 7500–7505 (2008)
12. Koonin, E.V.: Orthologs, paralogs, and evolutionary genomics. Annu. Rev. Genet. **39**, 309–338 (2005)
13. Overbeek, R., et al.: Use of contiguity on the chromosome to predict functional coupling. Silico Biol. **1**(2), 93–108 (1999)
14. Tatusov, R.L., et al.: The COG database: a tool for genome-scale analysis of protein functions and evolution. Nucleic Acids Res. **28**(1), 33–36 (2000)
15. Pellegrini, M., et al.: Assigning protein functions by comparative genome analysis: protein phylogenetic profiles. Proc. Natl. Acad. Sci. USA **96**(8), 4285–4288 (1999)
16. Date, S.V., Marcotte, E.M.: Discovery of uncharacterized cellular systems by genome-wide analysis of functional linkages. Nat. Biotechnol. **21**(9), 1055–1062 (2003)
17. Enault, F.: Annotation of bacterial genomes using improved phylogenomic profiles. Bioinformatics **19**(Suppl 1), i105–i107 (2003)
18. Altschul, S.F., et al.: Gapped BLAST and PSI-BLAST: a new generation of protein database search programs. Nucleic Acids Res. **25**(17), 3389–3402 (1997)
19. Kim, P.J. and N.D. Price, Genetic co-occurrence network across sequenced microbes. PLoS Comput Biol. 7(12): p. e1002340
20. Galperin, M.Y., Koonin, E.V.: Functional genomics and enzyme evolution. Homologous and analogous enzymes encoded in microbial genomes. Genetica **106**(1–2), 159–170 (1999)
21. Barker, D., Pagel, M.: Predicting functional gene links from phylogenetic-statistical analyses of whole genomes. PLoS Comput. Biol. **1**(1), e3 (2005)
22. Zheng, Y., Roberts, R.J., Kasif, S.:Genomic functional annotation using co-evolution profiles of gene clusters. Genome Biol. **3**(11), RESEARCH0060 (2002)
23. Sun, J., et al.: Refined phylogenetic profiles method for predicting protein–protein interactions. Bioinformatics **21**(16), 3409–3415 (2005)
24. Jothi, R., Przytycka, T.M., Aravind, L.: Discovering functional linkages and uncharacterized cellular pathways using phylogenetic profile comparisons: a comprehensive assessment. BMC Bioinform. **8**, 173 (2007)
25. Karimpour-Fard, A., Hunter, L., Gill, R.T.: Investigation of factors affecting prediction of protein–protein interaction networks by phylogenetic profiling. BMC Genomics **8**, 393 (2007)
26. Snitkin, E.S., et al.: Comparative assessment of performance and genome dependence among phylogenetic profiling methods. BMC Bioinform. **7**, 420 (2006)
27. Kim, Y., et al.: Inferring functional information from domain co-evolution. Bioinformatics **22**(1), 40–49 (2006)
28. Pazos, F., et al.: Correlated mutations contain information about protein–protein interaction. J. Mol. Biol. **271**(4), 511–523 (1997)
29. Gobel, U., et al.: Correlated mutations and residue contacts in proteins. Proteins **18**(4), 309–317 (1994)
30. Fodor, A.A., Aldrich, R.W.: Influence of conservation on calculations of amino acid covariance in multiple sequence alignments. Proteins **56**(2), 211–221 (2004)
31. Lockless, S.W., Ranganathan, R.: Evolutionarily conserved pathways of energetic connectivity in protein families. Science **286**(5438), 295–299 (1999)
32. Gloor, G.B., et al.: Mutual information in protein multiple sequence alignments reveals two classes of coevolving positions. Biochemistry **44**(19), 7156–7165 (2005)
33. Kann, M.G., et al.: Correlated evolution of interacting proteins: looking behind the mirrortree. J. Mol. Biol. **385**(1), 91–98 (2009)

34. Wang, Z.O., Pollock, D.D.: Coevolutionary patterns in cytochrome c oxidase subunit I depend on structural and functional context. J. Mol. Evol. **65**(5), 485–495 (2007)
35. Yeang, C.H., Haussler, D.: Detecting coevolution in and among protein domains. PLoS Comput. Biol. **3**(11), e211 (2007)
36. Burger, L., van Nimwegen, E.: Accurate prediction of protein–protein interactions from sequence alignments using a Bayesian method. Mol. Syst. Biol. **4**, 165 (2008)
37. Mintseris, J., Weng, Z.: Structure, function, and evolution of transient and obligate protein–protein interactions. Proc. Natl. Acad. Sci. USA **102**(31), 10930–10935 (2005)
38. Halperin, I., Wolfson, H., Nussinov, R.: Correlated mutations: advances and limitations. A study on fusion proteins and on the Cohesin–Dockerin families. Proteins **63**(4), 832–845 (2006)
39. Pazos, F., Valencia, A.: Protein co-evolution, co-adaptation and interactions. EMBO J. **27**(20), 2648–2655 (2008)
40. Ullsperger, C., Cozzarelli, N.R.: Contrasting enzymatic activities of topoisomerase IV and DNA gyrase from *Escherichia coli*. J. Biol. Chem. **271**(49), 31549–31555 (1996)
41. Weiss, D.S.: Bacterial cell division and the septal ring. Mol. Microbiol. **54**(3), 588–597 (2004)
42. Wang, X., Reyes-Lamothe, R., Sherratt, D.J.: Modulation of *Escherichia coli* sister chromosome cohesion by topoisomerase IV. Genes Dev. **22**(17), 2426–2433 (2008)
43. Fryxell, K.J.: The coevolution of gene family trees. Trends Genet. **12**(9), 364–369 (1996)
44. van Kesteren, R.E.: Co-evolution of ligand-receptor pairs in the vasopressin/oxytocin superfamily of bioactive peptides. J. Biol. Chem. **271**(7), 3619–3626 (1996)
45. Sato, T., et al.: The inference of protein-protein interactions by co-evolutionary analysis is improved by excluding the information about the phylogenetic relationships. Bioinformatics **21**(17), 3482–3489 (2005)
46. Goh, C.S., et al.: Co-evolution of proteins with their interaction partners. J. Mol. Biol. **299**(2), 283–293 (2000)
47. Pazos, F., Valencia, A.: Similarity of phylogenetic trees as indicator of protein–protein interaction. Protein Eng. **14**(9), 609–614 (2001)
48. Thompson, J.D., Gibson T.J., Higgins D.G.: Multiple sequence alignment using ClustalW and ClustalX. Curr Protoc Bioinform. **Chapter 2,** Unit 2 3 (2002)
49. Pazos, F., et al.: Assessing protein co-evolution in the context of the tree of life assists in the prediction of the interactome. J. Mol. Biol. **352**(4), 1002–1015 (2005)
50. Choi, K., Gomez, S.M.: Comparison of phylogenetic trees through alignment of embedded evolutionary distances. BMC Bioinform. **10**, 423 (2009)
51. Hakes, L., et al.: Specificity in protein interactions and its relationship with sequence diversity and coevolution. Proc. Natl. Acad. Sci. USA **104**(19), 7999–8004 (2007)
52. Muley, V.Y., Ranjan, A: Effect of Reference Genome Selection on the Performance of Computational Methods for Genome-wide Protein-Protein Interaction Prediction. PLoS ONE, In Press (2012)

Chapter 4
Chromosomal Proximity of Genes as an Indicator of Functional Linkage

Abstract Mostly prokaryotic genes have a tendency to be organized as clusters across chromosomes. Chromosomal proximity of genes, irrespective of the relative gene orientation, has been shown to be an indicative of their co-regulation. Genes that participate in related biological processes are often observed to be co-regulated. Hence, chromosomal proximity of genes has been proposed as a parameter indicative of functional linkages between them. However, prokaryotic genomes have been subjected to random rearrangements during evolution but these rearrangements are conservative in nature which invariably maintain individual genes in very specific functional and regulatory contexts. Hence, it is possible to deduce these rearrangements of genes based on chromosomal proximity of orthologous genes in multiple reference genomes. This chapter introduces the concept of genomic re-arrangements and discusses chromosomal proximity based three protein–protein interaction prediction methods.

4.1 Introduction

Mostly prokaryotic genomes are circular in nature and encoded genes have a tendency to be organized as clusters across chromosomes. As a contrast to the eukaryotic organisms, prokaryotes have compact genomes with a very few (if any) long overlaps between genes [1]. The adjacent genes within the clusters with same orientation (on the same genomic strand) often form co-transcriptional units called as operons [2, 3]. The operon, a group of co-transcribed and co-regulated genes, is one of the earliest and central concepts of bacterial genetics [4]. Co-transcription and co-regulation of a set of genes restrict their appearance or disappearance at the same time point in a cell. Thus, genes that are encoded from the same operon often perform related functions than genes that are not [2].

V. Y. Muley and V. Acharya, *Genome-Wide Prediction and Analysis of Protein–Protein Functional Linkages in Bacteria*, SpringerBriefs in Systems Biology, DOI: 10.1007/978-1-4614-4705-4_4,

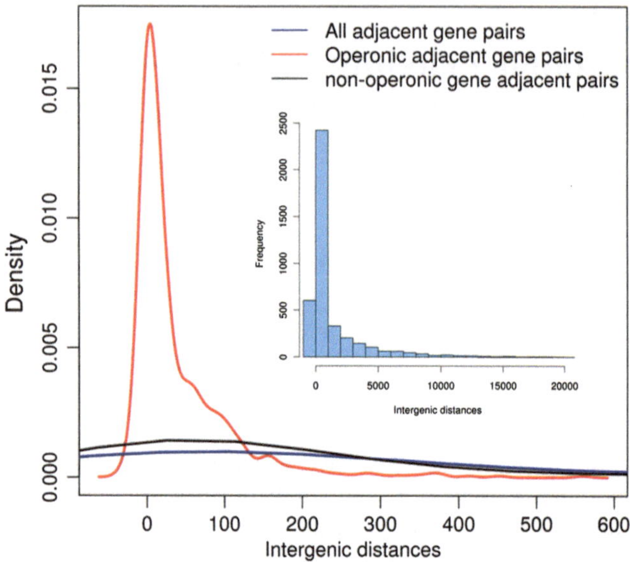

Fig. 4.1 Shorter intergenic distance between adjacent genes reflects their co-transcription. The figure was generated using adjacent genes of *Escherichia coli*. 666 operonic pairs used to as gold standard (*Red line*), and its known gene pairs that do not form operons (*Blue line*) whereas *black line* is for all adjacent gene pairs that are not operonic. Majority of operonic adjacent pairs have intergenic distances less than 100. Histogram shows most of the gene pairs are adjacent to each other (*Inset*). Operonic dataset obtained from EcoCyc database [6]

The rich knowledge of operon organization in prokaryotes has enabled researchers to perform an analysis of intergenic distances between the genes. As shown in Fig. 4.1, the frequency distribution of intergenic distances between genes in the same operons shows clear peak at short distances, as contrasts with a flat frequency distribution of genes that are adjacent but not in the same operons [2]. It is also observed that genes in the same operon tend to be involved in the same biological function. These observations led to implement a method to predict the genomic organization of genes into transcription units, based on short intergenic distance between them, with a maximum accuracy of 88 % in *Escherichia coli* genome [2]. This step opened up the possibility of predicting functional coupling using contiguity of genes on the chromosome in prokaryotes whose genome sequences have been finished. It is because, on an average 35 % of genes that are part of various clusters acts in various metabolic pathways [5]. Thereby, genes that are located within short intergenic distances with same orientation (order) on the chromosome are likely to co-transcribed, co-regulated, and hence likely to be involved in the same biological process.

Since biology is full of surprise and uncertainty, operonic organization is not an exception to the rule. During the evolution of prokaryotic genomes, though operons are highly conserved, yet no conservation is seen in large-scale synteny or gene order [1, 7]. It means that during evolution genes are well conserved but orientation or order of genes within the operon is not maintained. Although, using comparative genomic analysis, few operons have been revealed with conserved gene orders that are shared by a broad range of organisms [8, 9]. As observed earlier, products of these highly conserved operons typically interact with each other physically [10], a trend which reflects selection against the deleterious effects of imbalance between protein complex subunits [1, 11]. The most dramatic example of this trend is the ribosomal super-operon that includes over 50 genes. In the 1990s, it was hypothesized that the ribosomal gene cluster in the 'last common universal ancestor' was broken up into smaller clusters in the course of evolution. The follow-up analysis revealed more complex evolutionary scenario for ribosomal gene cluster. The large ancestral cluster is not only broken up during evolution but also involved in the joining of clusters, break up into further smaller clusters, and the rearrangement of these into new clusters [8, 12]. There are several lines of evidence on the conservative nature of these rearrangements that invariably maintained individual genes in very specific functional and regulatory contexts [12, 13]. This conserved context led to the notion of an uber-operon [12] or a conserved gene neighborhood [13] that represents an array of overlapping, partially conserved operons (known or predicted) present in a prokaryotic genomes. The majority of genes in the uber-operons encode proteins that participate in the same pathway and/or complex.

For example, the neighborhood of ParC and ParE proteins is depicted in Fig. 4.2. These two proteins form active topoisomerase IV complex and belong to the type II topoisomerase family. Topoisomerase IV plays an essential role by removing double-stranded DNA crossings while progression of the replication fork and the chromosome segregation after replication [14, 15]. These proteins are not operonic or adjacent in *E. coli* and in many other organisms. However, as shown in Fig. 4.2, these two proteins are probably encoded from the same operon in *Haeamophilus influenzae*. There are also number of other organisms in which these two proteins are adjacent in genomes. Therefore, even if they are not operonic in *E. coli* and in many other species, the evidence of their conserved chromosomal proximity in other genomes is enough to predict their functional linkage. Furthermore, in some of the organisms, ParC is proximal to the MukB, GyrA, GyrB, and InfB. Functional relevance of these proteins is consistent with the role played by ParC in chromosome segregation. MukB is known to play a central role in chromosome condensation and segregation, whereas GyrA and GyrB are known to be involved in topoisomerase activity. The neighborhood context shown in Fig. 4.2 is inferred for *E. coli* protein ParC by searching its orthologs in the represented genomes. There are several other examples of operon re-associations that have been reported in the literature such as predicted overlapping operons that encode subunits of the archaeal exosomal complex [16], and an antivirus defense system encoded by cas genes [13, 17, 18]. Along with the genes with related functions, highly conserved arrangements also include genes with apparently unrelated functions. For example, the common occurrence of proteosome subunit genes in the archeal exosome neighborhood and the enolase gene in the operons encoding ribosomal genes [1].

The presence of related genes in the conserved neighborhood suggests a possible scenario of 'purifying selection' where the separation of these genes would be

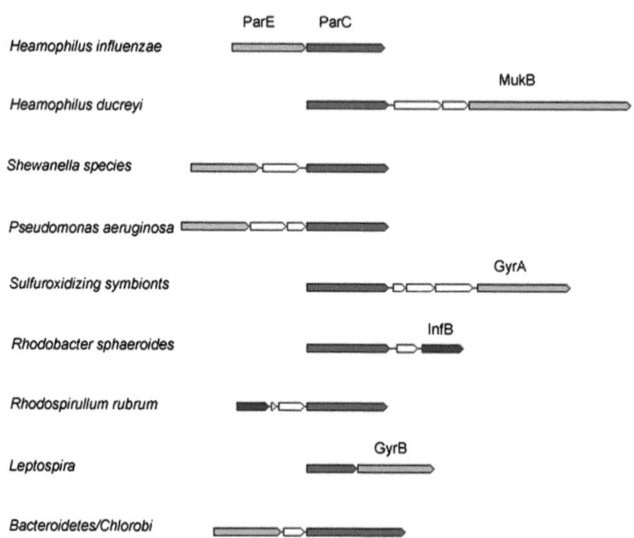

Fig. 4.2 A genomic neighborhood of ParC protein. STRING [19] database used to derive neighborhood context of *Escherichia coli* ParC. Although *E. coli* ParE and ParC are localized away from each other, many organisms show evidence of their co-regulation. The combination of ParC and ParE proteins plays important role in DNA segregation and replication. Even closely related species such as *Haeamophilus* do not share these two genes but *Pseudomonas* and *Shewanella* species does

selectively disadvantageous and would thus be eliminated from the population [1]. Conversely, the genes that are placed within new clusters of functionally and regulatory related genes during rearrangements would have a relatively small impact on the fitness of the organism and could be maintained through drift and fixation [1]. The presence of seemingly unrelated genes in the conserved neighborhood is computationally difficult to understand but can be possibly explained as,

- a case of 'gene sharing' i.e., multiple functional roles played by respective genes [1];
- a case of "genomic hitchhiking", where genes with different functions maintained in the neighborhood during evolution due to requirements at the same time in the cell. Hence, the conserved operonic organization of such genes helps them to express at the same time point [13, 20].

The exception to the great majority of the co-directional conserved gene neighborhoods in prokaryotes [21] is the conserved adjacent bidirectionally transcribed genes, i.e., 'divergently' organized coding regions [22]. It has been shown that these conserved gene pairs are strongly co-regulated by virtue of bidirectional transcription from symmetric promoters, and are functionally associated [22].

All previous analyses suggest that the genome organization in prokaryotes is not uniform due to the frequently occurring dynamic re-arrangements [1, 7, 23].

The task of identifying such re-arrange operons or divergent transcriptional units is relatively simple yet very powerful. Furthermore, it provides plenty of opportunity to point out functional roles of the uncharacterized proteins. With the available genome sequences, it will be interesting to know how we can use chromosomal proximity or neighborhood of genes to infer functional linkage among them.

4.2 Occurrence of Genes in the Operon as an Indicator of Functional Linkage

With intergenic distances between co-directional genes as a sole criterion, operons can be predicted exceptionally well with estimated accuracy of 88 %. Gene products that are encoded from the same operon often perform related biological functions. Therefore, co-presence in operon is the simplest available option to predict functional linkages among genes. Operon prediction task is a simple three-step approach as explained below.

- First, we need information about the gene coordinates on genome sequence for an organism of interest. The protein table files with extension 'ptt' usually contain gene coordinate information. These files can be downloaded from NCBI ftp (ftp://ftp.ncbi.nih.gov/genomes/Bacteria/).
- In the second step, a simple program can be written to calculate intergenic distance between adjacent gene pairs based on their coordinates from protein table file.
  - These distances should be calculated for gene pairs on plus and minus genomic strand independently.
- Finally, the frequency distributions of intergenic distances for gene pairs can be plotted known to be operonic and nonoperonic. Then, we will choose distance as a cutoff value, where the highest peak is observed which become stable and flat at higher intergenic distances between gene pairs. Below this intergenic distance cutoff, we have probable operonic pairs, which can be used to assemble adjacent gene pairs into clusters or complete operons. This is exemplified in Fig. 4.1 in which the peak is near intergenic distance cutoff of 100 nucleotide bases.
  - In case, if operons are not known for the organism of interest, then arbitrary cutoff below 200 can be considered as a best way to determine probable operonic pairs. Lesser the cutoff better the operon predictions.
  - If one is interested in large-scale analyses then one can follow the above-mentioned strategy, otherwise more sophisticated approaches are available for operon predictions [24–28].
  - Most of these methods use various features to improve the quality of predictions such as gene expression data, clustering of orthologous gene pairs, phylogenetic profiling, position of transcription terminator sites, etc. [24–30].

One can assume functional linkages between the genes that co-presence in the predicted operons, since these adjacent gene pairs are co-directional and are likely to co-transcribe.

4.3 Co-occurrence of Orthologous Gene Pairs in the Same Operon as an Indicator of Functional Linkage

The use of above-mentioned method to infer functional linkages is limited only to those genes that are adjacent on the chromosome. That is, for a query genome with N genes, operon prediction can generate scores for at most N gene pairs from the N^2 possible pairs. Therefore, co-regulated and functionally linked genes, if placed away from each other on the chromosome during re-arrangements in the course of evolution, will not be detected by simple operon prediction approach.

Comparative genomic approach has become a powerful tool to deduce the re-arrangements of genes based on chromosomal proximity of orthologous genes. This brings out a better version of operon prediction methods called as Gene Cluster (GC), in which GCs can be defined (but not in the query genome) as sets of co-directional genes within intergenic distance threshold of certain nucleotide bases in all reference genomes (Fig. 4.3a). Then, the GC algorithm considers every possible pair of gene products encoded from a query genome and calculates frequency with which coding genes of their orthologs co-occur in the same gene cluster in reference genomes [5, 10, 26, 31]. GC scores above zero indicate the co-presence of genes encoding orthologs of query protein pair in the same operon at least in one reference genome. Therefore, GC is likely to discover operons that are re-arranged in the query genome, based on the evidence of their intact operon structure in multiple reference genomes. Hence, even if query gene pair is no longer proximal on the chromosome, one can speculate their possible co-regulation and thereby functional coupling as evident from the reference genomes.

However, GC algorithm also suffers from their limitation as GC gathers evidences from the limited number of orthologous genes that are co-directional and proximal in the reference genomes. Therefore, the prediction coverage of GC is not good as compared to existing methods. Furthermore, this method will not able to identify re-arranged operons that are divergently transcribed, often co-regulated, and functionally linked [22].

4.4 Genomic Neighborhood as an Indicator of Functional Linkage

One slightly modified form of GC algorithm referred to as Gene Neighbor (GN) method, which overcome limitations of GC by not only predicting re-arranged

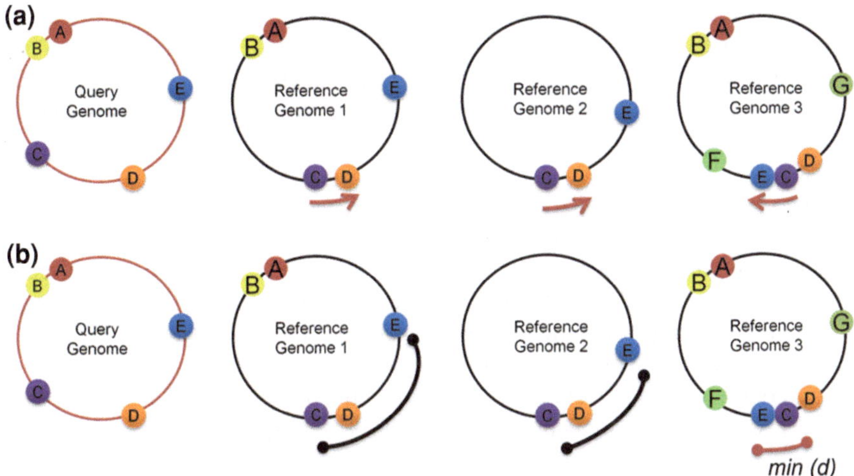

Fig. 4.3 A schematic representation of gene cluster and gene neighbor method for predicting protein–protein interactions. **a** Gene Cluster calculates co-occurrence probability of orthologs of query proteins encoded from the same gene clusters in reference genomes. Gene clusters are defined as a set of unidirectional genes within intergenic distance of 100 nucleotide bases. In the given example, genes encoding orthologs of query proteins C and D co-occur in two reference genomes, hence 2/3 is interaction score between them. **b** Gene Neighbor method calculates interaction scores for query protein pairs based on the minimum chromosomal distance between their orthologs encoding genes in any one of the reference genome. In the given example, minimum distance for proteins C and E is obtained from third reference genome and that would be the interaction score for protein C and E of query genome. Gene Cluster calculates co-occurrence probability of orthologs of query proteins encoded from same gene clusters in reference genomes. Gene clusters are defined as a set of unidirectional genes within intergenic distance of 100 nucleotide bases. In the given example, genes encoding orthologs of query proteins C and D co-occur in two reference genomes, hence 2/3 is the interaction score between them

operons but also divergently transcribed gene pairs. GN assumes chromosomal proximity of orthologous genes, irrespective of relative gene orientations, across a set of reference genomes as an indicator of functional linkage [22, 32].

Over the years, GN method has been modified into several forms [5, 10, 22, 26, 31–34]. The most used form of GN method considered two genes from the query genome, lets say X and Y. Then the distance D_i for these genes obtained from reference genome 'i' by calculating the chromosomal distance between the orthologous genes of X and Y. Since the bacterial genomes are mostly circular in nature, the distance D_i has to be calculated in both clockwise and anti-clockwise direction. The minimum of these two values normalized by the chromosome length of reference genome 'i' is the final distance D_i between query genes X and Y. The mathematical formula to calculate the GN score is as follows,

$$D_i = \frac{2d}{N}$$

where, d is the chromosomal distance between orthologs of gene X and Y in the ith reference genome. N is a chromosome length of reference genome i.

In order to minimize the effect of larger intergenic distances between genes, the distance can be computed as the number of genes that appear between the orthologs of genes X and Y plus 1 (hence adjacent genes have a distance of 1), instead of computing chromosomal distance between orthologous genes. Then this distance has to be normalized by total number of genes of an organism. Afterwards, overall procedure to calculate D_i is same as mentioned above.

The minimum distances calculated across reference genomes can be used to calculate joint probability (GN interaction score for a particular gene pair) that the distances are smaller than the observed distances. Recent analysis has suggested that the minimum distance of orthologous genes on the chromosome of any one of the reference genome is enough to infer functional linkage between genes say X and Y with average prediction accuracy of 89 % [26] (Fig. 4.3b).

4.5 Scope and Future Perspective

We have discussed three forms of chromosomal proximity-based methods, i.e., Operon, GC, and GN for the identification of re-arranged, co-directional, and divergent transcriptional units. These methods have been highly studied and several other features have been used along with intergenic distance as a major one. These features include, gene expression, ribosomal binding sites, terminator sites, and so on. One of the study based on a comparison of the performance of operon predictions on *E. coli* and *Bacillus subtilis* suggests that there is still room for improvement in the predictions [24]. In our opinion, if there is any margin to improve operon prediction, one can possibly consider the composition of DNA sequences of adjacent genes as well as the consideration of short stretches of upstream and downstream regions. Furthermore, the physicochemical properties of DNA sequences can also be used to enhance the prediction ability over existing approaches. However, the question is to what extent these features will improve prediction accuracy. The latest analyses have reported accuracy of 93 % [35].

References

1. Koonin, E.V.: Evolution of genome architecture. Int. J. Biochem. Cell Biol. **41**(2), 298–306 (2009)
2. Salgado, H.: Operons in *Escherichia coli*: genomic analyses and predictions. Proc. Natl. Acad. Sci. USA **97**(12), 6652–6657 (2000)
3. Beckwith, J.: The operon as paradigm: normal science and the beginning of biological complexity. J. Mol. Biol. **409**(1), 7–13 (2011)
4. Jacob, F., Monod, J.: Genetic regulatory mechanisms in the synthesis of proteins. J. Mol. Biol. **3**, 318–356 (1961)

5. Overbeek, R., et al.: The use of gene clusters to infer functional coupling. Proc. Natl. Acad. Sci. USA **96**(6), 2896–2901 (1999)
6. Keseler, I.M., et al.: EcoCyc: a comprehensive view of *Escherichia coli* biology. Nucleic Acids Res. **37**(Database issue), D464–D470 (2009)
7. Mushegian, A.R., Koonin, E.V.: Gene order is not conserved in bacterial evolution. Trends Genet. **12**(8), 289–290 (1996)
8. Wolf, Y.I.: Genome alignment, evolution of prokaryotic genome organization, and prediction of gene function using genomic context. Genome Res. **11**(3), 356–372 (2001)
9. Itoh, T., et al.: Evolutionary instability of operon structures disclosed by sequence comparisons of complete microbial genomes. Mol. Biol. Evol. **16**(3), 332–346 (1999)
10. Dandekar, T., et al.: Conservation of gene order: a fingerprint of proteins that physically interact. Trends Biochem. Sci. **23**(9), 324–328 (1998)
11. Papp, B., Pal, C., Hurst, L.D.: Dosage sensitivity and the evolution of gene families in yeast. Nature **424**(6945), 194–197 (2003)
12. Lathe 3rd, W.C., Snel, B., Bork, P.: Gene context conservation of a higher order than operons. Trends Biochem. Sci. **25**(10), 474–479 (2000)
13. Rogozin, I.B., et al.: Connected gene neighborhoods in prokaryotic genomes. Nucleic Acids Res. **30**(10), 2212–2223 (2002)
14. Ullsperger, C., Cozzarelli, N.R.: Contrasting enzymatic activities of topoisomerase IV and DNA gyrase from *Escherichia coli*. J. Biol. Chem. **271**(49), 31549–31555 (1996)
15. Weiss, D.S.: Bacterial cell division and the septal ring. Mol. Microbiol. **54**(3), 588–597 (2004)
16. Koonin, E.V., Wolf, Y.I., Aravind, L.: Prediction of the archaeal exosome and its connections with the proteasome and the translation and transcription machineries by a comparative-genomic approach. Genome Res. **11**(2), 240–252 (2001)
17. Makarova, K.S., et al.: Defense islands in bacterial and archaeal genomes and prediction of novel defense systems. J Bacteriol. **193**(21), 6039–6056 (2011)
18. Makarova, K.S., et al.: Evolution and classification of the CRISPR-Cas systems. Nat. Rev. Microbiol. **9**(6), 467–477 (2011)
19. Jensen, L.J., et al.: STRING 8—a global view on proteins and their functional interactions in 630 organisms. Nucleic Acids Res. **37**(Database issue), D412–D416 (2009)
20. Rogozin, I.B., et al.: Purifying and directional selection in overlapping prokaryotic genes. Trends Genet. **18**(5), 228–232 (2002)
21. Rogozin, I.B., et al.: Congruent evolution of different classes of non-coding DNA in prokaryotic genomes. Nucleic Acids Res. **30**(19), 4264–4271 (2002)
22. Korbel, J.O., et al.: Analysis of genomic context: prediction of functional associations from conserved bidirectionally transcribed gene pairs. Nat. Biotechnol. **22**(7), 911–917 (2004)
23. Watanabe, H., et al.: Genome plasticity as a paradigm of eubacteria evolution. J. Mol. Evol. **44**(Suppl 1), S57–S64 (1997)
24. Brouwer, R.W., Kuipers, O.P., van Hijum, S.A.: The relative value of operon predictions. Brief Bioinform. **9**(5), 367–375 (2008)
25. Price, M.N., et al.: A novel method for accurate operon predictions in all sequenced prokaryotes. Nucleic Acids Res. **33**(3), 880–892 (2005)
26. Yellaboina, S., Goyal, K., Mande, S.C.: Inferring genome-wide functional linkages in *E. coli* by combining improved genome context methods: comparison with high-throughput experimental data. Genome Res. **17**(4), 527–535 (2007)
27. Janga, S.C., et al.: The distinctive signatures of promoter regions and operon junctions across prokaryotes. Nucleic Acids Res. **34**(14), 3980–3987 (2006)
28. Moreno-Hagelsieb, G., Collado-Vides, J.: A powerful non-homology method for the prediction of operons in prokaryotes. Bioinformatics **18**(Suppl 1), S329–S336 (2002)
29. Ranjan, S., Gundu, R.K., Ranjan, A.: MycoperonDB: a database of computationally identified operons and transcriptional units in Mycobacteria. BMC Bioinform. **7**(Suppl 5), S9 (2006)
30. Bergman, N.H., et al.: Operon prediction for sequenced bacterial genomes without experimental information. Appl. Environ. Microbiol. **73**(3), 846–854 (2007)

31. Janga, S.C., Collado-Vides, J., Moreno-Hagelsieb, G.: Nebulon: a system for the inference of functional relationships of gene products from the rearrangement of predicted operons. Nucleic Acids Res. **33**(8), 2521–2530 (2005)
32. Tamames, J., et al.: Conserved clusters of functionally related genes in two bacterial genomes. J. Mol. Evol. **44**(1), 66–73 (1997)
33. Bowers, P.M., et al.: Prolinks: a database of protein functional linkages derived from coevolution. Genome Biol. **5**(5), R35 (2004)
34. Ferrer, L., Dale, J.M., Karp, P.D.: A systematic study of genome context methods: calibration, normalization and combination. BMC Bioinform. **11**, 493 (2010)
35. Bockhorst, J., et al.: Predicting bacterial transcription units using sequence and expression data. Bioinformatics **19**(Suppl 1), i34–i43 (2003)

Chapter 5
Analyses of Complex Genome-Scale Biological Networks

Abstract Cellular systems are organized as a complex web of interactions among numerous macromolecules. Among the others, proteins are important since they play important role in virtually every biological process that occurs in the cell. Cellular systems are constantly challenged by fluctuations in the surrounding environment. In response, repertoire of the protein contents in the cell constantly alters, accordingly the interactions among them. Mathematically, these protein–protein interactions (PPIs) can be conceptualized in the form of graph or network for ease in analysis. A node in the graph represents protein and its link with other node is represented by edge. The local and global topological properties of the network reveal organization principles of underlying interactions among total proteins of an organism. The local properties specify importance of a particular protein in the network whereas global properties reflect their organization operational in the cell. Over the years, several graph theoretic and clustering techniques proposed for analysis of complex physical world have been applied to understand dynamic organization of the cellular networks. These methods promise to become more informative as the high quality PPI networks increase by orders of magnitude. This chapter provides an overview on various topological properties of networks and their significance in understanding biological systems.

5.1 Introduction

Unknowingly networks are part of every aspect of our life. We are encompassed by complex systems that are made up of small components interconnected at various levels with each other. From our circle of friend, Internet connections, transportation, government organization, to that of computer chips, everything is connected within these systems. Biological systems are not exception to these rules. The cells are composed of components that are connected with each other

V. Y. Muley and V. Acharya, *Genome-Wide Prediction and Analysis of Protein–Protein Functional Linkages in Bacteria*, SpringerBriefs in Systems Biology, DOI: 10.1007/978-1-4614-4705-4_5,

and interplay of neurons in brain to control our behavior is also governed by fine-tuning of interactions among them. In order to steer and control we should understand the organization of these systems in detail. At the end of the twentieth century, two scientific papers were published by Albert Barabasi group which had stirred the physical and biological sciences [1, 2]. In the first paper, Barabasi and coworkers suggested that many large networks including the network of hyperlinks in the World Wide Web (WWW) are not randomly organized but are scale-free [1]. It implies that the connectivity of nodes in the networks follows a scale-free power-law distribution. In the second paper, they analyzed metabolic networks of 43 organisms representing all three domains of life [2]. They observed metabolic networks have the same topological properties, which were observed for complex non-biological systems. These two studies have opened new avenues to reveal universal organizational principles underlying the complex systems. Over the years, network analyses have witnessed significant progress in almost every aspect of life such as biology, economy, power grids, social world, transportation, etc. [1, 3]. Several properties of biological systems could be explained using network/graph theory such as transcriptional regulation, signaling, protein–protein interaction(s) (PPI), etc. In this section, we discuss some of the important topological properties useful for dissecting complex PPI networks in order to understand characteristics and organization of cellular systems. The network properties we discussed here are limited to the undirected networks since PPI networks have been represented as binary interactions where A–B is considered similar to that of B–A.

5.2 Network Representation of Biological Systems

A graph is represented as points and the lines connecting them. Each point is called as a node or a vertex and the line connecting two nodes is called as an edge. In the PPI network, proteins are nodes and interactions between them are edges (Fig. 5.1). The connections between numerous proteins forms a web of interaction, which is termed as network; mathematically it is a graph. The original idea of network came from social network at first and then from computer network, which was further implicated in biological network. The arrangement of nodes (proteins) and edges (interactions) all together result into a characteristic topology of network. Not all nodes are equally important in the network. Some nodes are more critically involved in the maintenance and integrity of network than the others. There are various ways of retrieving topologically important constituents of the network, which help us to analyze the structure of networks at various levels of organization. Advances in the network field allows us to understand network structure at three major levels which are as follows:

1. Organization of individual nodes (or proteins) in the network.
2. Organization of groups of nodes (or proteins) in the network.
3. Organization of the whole network.

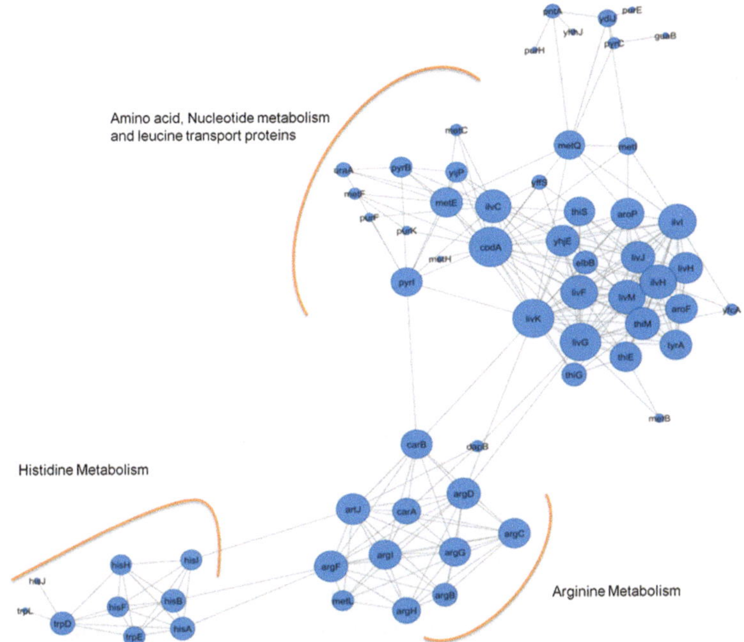

Fig. 5.1 A representation of protein–protein interaction network. The size of each node is proportional to its number of interacting partners. CodA protein has the highest number of interacting partners, which act in the pyrimidine salvage pathway. CodA, LivG, LivF, and LivK proteins can be categorized as hubs in the network due to their higher number of connections. Network can be divided into three components based on interconnectivity of proteins with each other. Histidine and arginine metabolic pathway proteins show higher number of connections among them than rest of the proteins in the network. The network is derived from co-expression of these proteins in various physiological conditions. Microarray data were used from M3D database [4]

5.3 Organization of Individual Nodes in the Network

There are the properties of each node of the network, which defines its local structure and positioning in network. The following are the measures that calculate local topological properties of the nodes.

5.3.1 Degree Centrality

The most basic structural property of a protein is its degree (or connectivity), k, which is number of connections/links the protein has to other proteins in the network. The degree (k) of a particular protein reflects its importance in the network

[5] and proteins with higher k value, called as hubs (Fig. 5.1). Proteins that act as hubs in the network are often essential for cell survival since they can distort the structure of network when deleted [6]. According to the previous reports, an average degree of 2–10 was estimated in a typical functioning cell [7, 8].

For a given node v, the degree centrality (C_d) is defined as,

$$C_d(v) = n(v)$$

where, n is the number of connections of node v in the network.

5.3.2 Hubs

Hubs are highly connected nodes of a network. Compared to any random nodes, they are often found to be more essential for maintaining the overall topology of the network. These proteins could either use single or multiple interfaces for binding to its partners. The number of interfaces used for connection with partners also depends on the co-expression of the hub and its partners. If the partners involved in interaction are co-expressed all together with the hub then the hub is known as *party hub* and it uses multiple-interfaces for the same. Single-interface hubs interact with one protein at a time, so the partners may not be expressed together and thus, these kinds of hubs are known as *date hubs*. Hub proteins with one or two binding interfaces tend to be more disordered than other proteins [9]. The disordered residues in multi-interface hubs are comparable to that of overall proteome. However, binding interfaces in single-interface or multi-interface hub proteins are highly structured [9].

There have been numerous studies on the functional role of these hub proteins in cellular network [6, 10]. Mostly hub proteins are found to be involved in molecular function of regulation and complex formation [6, 10, 11]. The higher level of disorder of single-interface hubs is also implicated in a cascade formation through binding with each other during events like signaling pathways [9].

5.3.3 Clustering Coefficient

Clustering coefficient (CC) of a node in the network measures the interconnections of its adjacent neighbors, which is defined as follows [3],

$$\mathrm{CC}(v) = \frac{2n_{vi}}{k(k-1)}$$

where, n_{vi} is the number of links connecting the k neighbors of node v to each other. It ranges on the scale, 0–1, score of one reflects all the neighbors of node under investigation are interconnected whereas zero reflects no connections between them.

The CC of a network is the average CC of all nodes in the network [3]. The higher CC of a protein reflects interconnectivity of its partners, hence implies the ability of these proteins to occupy the same subsystem and likelihood of functional association. Moreover, the higher CC values reflect that any external signal is likely to flow within the subsystem and hence less efficiently propagates in a whole network.

5.3.4 Shortest Path and Mean Path Length

Distance between two nodes is measured with the *path length*, which is the number of links/edges we need to pass through to travel between them. For given any two random nodes of a network, there can be number of possible paths connecting those two nodes, the path with the smallest number of links between the selected nodes is often considered for analysis and is called as *shortest path*. The average or characteristic path length represents the average of shortest paths between all pairs of nodes in the network. The average path length represents the network's overall navigability [3, 5].

Previous analyses on the network properties have suggested that networks can be highly clustered (i.e., higher average CC), like a regular graph, yet can have small characteristic path lengths similar to a random graph [3].

5.3.5 Closeness Centrality

Closeness centrality of a node quantifies its closeness to the other nodes in the network, which is defined as follows [3],

$$\text{Closeness}(v) = \sum\nolimits_j [d_{vj}]^{-1} = \frac{1}{\sum_j d_{vj}}$$

where, v is the focal node, j is another node in the network, and d_{vj} is the shortest distance between these number nodes. The proteins with high closeness values in the networks are typically close to and can communicate quickly with the other nodes in the network.

5.3.6 Betweenness Centrality

Betweenness is one of the most important local properties which measure the number of shortest paths going through a certain node. It is calculated as follows,

$$\text{Betweenness}(v) = \sum\nolimits_{ij} \frac{p_{ij}(v)}{p_{ij}}$$

where, p_{ij} is the number of shortest paths between nodes i and j. $p_{ij}(v)$ is the number of shortest paths between nodes i and j going through node v.

Nodes with high betweenness values represent critical points in the network, which are called as *bottlenecks* of the network [12]. These are analogous to major bridges and tunnels on a highway map connecting two lands separated by mountains or rivers. There are two types of nodes that are important, which show high betweenness values. First, a node is a hub and is connected with another hub in the network. Second, a node with very less number of connections which connects two hubs. The latter types of nodes in the network are often called as *articulation points*. The random removal of such node can break the communication between two hubs and hence, it can isolate two subnetworks. It has been observed that nodes with high betweenness are good predictor of essentiality [12].

5.4 Organization of the Whole Network

There are several ways to analyze topological properties of networks, which reveal global organization. As shown in Fig. 5.2, global organization of the nodes in the PPI network and the genetic network is quite different. However, simply looking at the figures of these two networks we know little about the organization of various nodes in them. We describe some of the important properties, which dissect these networks at global level in the subsequent context.

5.4.1 Random Networks

A random network model proposed by Erdös–Rényi (ER) starts with N nodes and connects each pair of nodes with probability p, which creates a graph with approximately pN(N–1)/2 randomly placed links [5]. The degree of all the nodes in a network approximates a Poisson distribution, which indicates that most nodes have approximately the same connectivity (close to the average degree $<k>$). The tail (high k region) of the degree distribution $P(k)$ decreases exponentially, which indicates that nodes that significantly deviate from the average are extremely rare. The CC is not correlated with node's degree. The average path length is proportional to the logarithm of the network size, $l \sim \log N$, which indicates that it is characterized by the small world property.

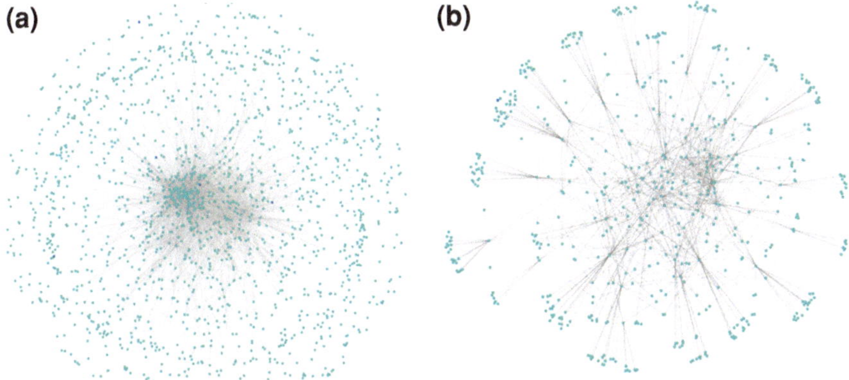

Fig. 5.2 A global organization of biological networks. **a** Protein–protein interaction network, which is reconstructed by combining two experimental dataset [13, 14]. It consists of 7,277 interacting pairs among 1,644 nodes. **b** Genetic interaction network, which is reconstructed using synthetic lethal gene pairs [15]. It consists of 1,171 interacting pairs among 576 protein-coding genes

5.4.2 Biological Networks are Scale-Free

Scale-free networks are characterized by a power-law degree distribution. Biological networks follow power-law degree distribution and hence are scale-free in nature. The probability that a node has k links follows $P(k) \sim k$–γ, where γ is the degree exponent. The value of γ determines many properties of a network. The role of the hubs becomes important in the network when values of γ are smaller while for $\gamma > 3$, the importance of hubs diminishes. For $2 < \gamma < 3$, there is a hierarchy of hubs, with the most connected hub being in contact with a small fraction of all nodes [5]. Therefore, scale-free networks have an inherent robustness against random node failures, although they are sensitive to the failure of hubs [6]. The degree exponent range for most of the biological and non-biological networks is $2 < \gamma < 3$. The biological networks are ultra-small, with the average path length (l) following $l \sim \log(\log(N))$, which is significantly shorter than $\log(N)$ that characterizes random small-world networks [16, 17].

According to the Barabási–Albert model, the highly connected nodes in these scale-free networks are statistically significant than in a random graph, although the network topology is determined by only a small fraction of hubs. The probability with which newly added node connects to the existing nodes of the network is directly proportional to the connectivity of the existing nodes, resulting in a phenomenon called '*rich get richer*' or *preferential attachment* [1]. Despite these efforts whether PPI networks follow power-law degree distribution or not is under debate [18, 19].

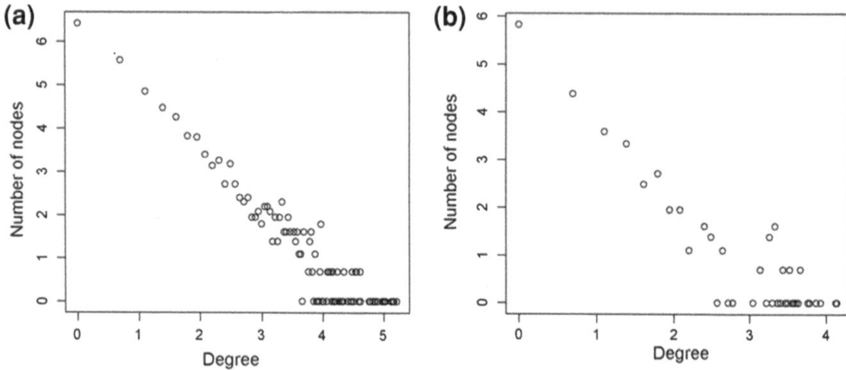

Fig. 5.3 Log–log plots of a degree distribution. **a** The degree distribution of protein–protein interaction network. **b** The degree distribution of genetic interaction networks. These distributions clearly show that both networks follow power-law where few nodes with higher number of degrees

5.4.3 Degree Distribution

The degree distribution, $P(k)$, suggests the probability that a selected protein has exactly k interacting partners. $P(k)$ is obtained by counting the number of proteins $N(k)$ with $k = 1, 2, \ldots$ links and dividing by the total number of proteins in the network, i.e., N. The degree of distribution forms basis to distinguish between different classes of networks. The degree of distribution of PPI and genetic interaction networks is depicted in the Fig. 5.3, which indicates that a few hubs/proteins hold together numerous proteins. Such distributions are defined as power-law degree distribution in contrast to the distribution of random networks where the system has a characteristic degree and there are no highly connected nodes (or hubs).

5.4.4 Assortativity and Disassortativity

We have already discussed that biological networks are sparsed with a very few hub proteins. An interesting question is whether hubs are more prone to be connected to other hubs. As shown in Fig. 5.4, the networks can be classified as *assortative*, if hub proteins tend to be connected to other hubs (Fig. 5.4a). In case, hub proteins are prone to be connected to other hubs via nodes with low degree then the networks are called as *disassortative* (Fig. 5.4b).

The biological networks were believed to be disassortative, where chance of failure of two hubs simultaneously is lower since they are not directly connected

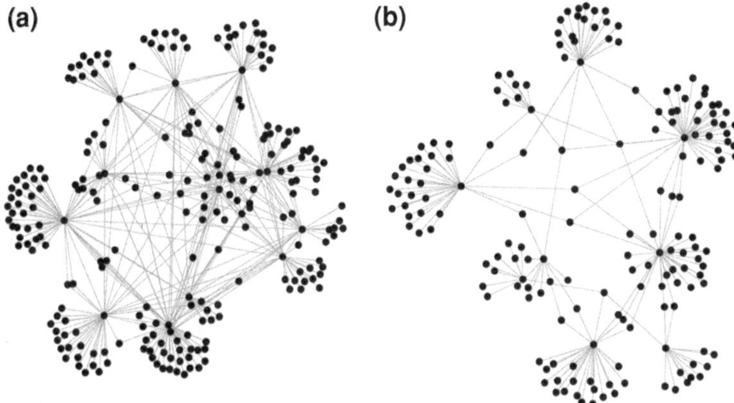

Fig. 5.4 A schematic illustration of assortative and disassortative networks. **a** Assortative network, where hubs are connected to other hub proteins. **b** Disassortative network, where hubs are connected to other hubs via nodes with small number of connections. In this network, the nodes connecting two hubs are articulation points (or nodes with high betweenness)

with each other. Thereby, this organization of the hubs in the network provides specificity of functional modules along with the stability of networks [20].

Assortativity and disassortativity of the networks can be quantified by positive and negative correlation of degree of nodes and the average connectivity of their neighbors, respectively. A negative correlation between average neighborhood connectivity and the connectivity (degree) was observed for full yeast PPI dataset indicating that highly connected proteins tend to be isolated from each other [20] as opposed to the highly connected nodes that tend to interact with other highly connected nodes [5, 21]. The highly filtered yeast data of PPIs that were characterized by at least two experiments and three experiments show slightly positive correlation of degree and average neighborhood connectivity [21]. However, a recent analysis suggested that biological networks have characteristics of both types of properties [22]. We have plotted distribution of node degrees and the average connectivity of their neighbors for PPI and genetic interaction network (Fig. 5.5). It seems that the PPI network consists of both types of properties (Fig. 5.5a), whereas the genetic network consists of hubs that are not connected with each other (Fig. 5.5b). This difference between these two networks suggests that the genetic networks are more robust than the PPI networks.

5.4.5 Network Diameter

The network diameter is defined as the maximum length of shortest paths between two nodes of the largest connected component of a network [3]. It is also defined as the average shortest path length between all pairs of nodes in a network [23]. Biological networks and real-world networks are known to have the small-world

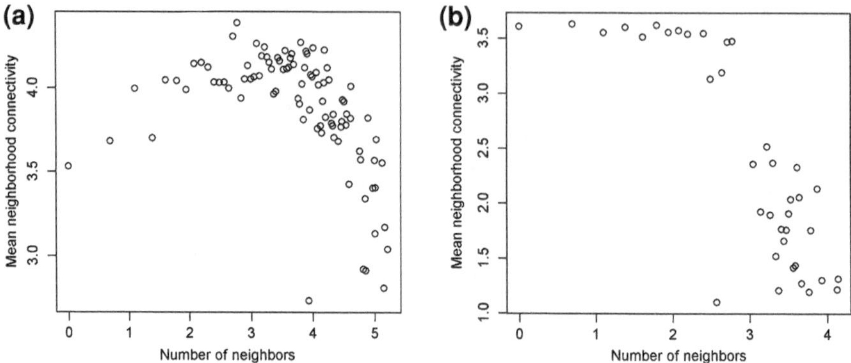

Fig. 5.5 Log–log plot of degree versus average neighborhood connectivity. **a** Protein–protein interaction network (PIN). **b** Genetic Interaction Network (GIN). There is a weak positive correlation between degree and connectivity of neighbors for PIN, which drops quickly at higher degree nodes and becomes negative. It suggests that the network consists of hubs that are connected to other hubs along with hubs that are connected via low degree nodes. The distribution of degree and average neighborhood connectivity for GIN suggests that the hubs are not directly connected to each other

property which is characterized by the small diameter of a network [3]. The networks with small-world architecture are known to be highly efficient in exchanging information and may also serve to minimize transition between nodes [24, 25]. However, a recent study suggests that the short diameters of real-world networks may be a consequence of higher modularity [23]. Thereby, shorter diameters could provide higher functional efficiency to a network. Furthermore, their result suggests a tradeoff between network efficiency and multi-functionality, robustness, and/or evolvability.

5.5 Organization of Groups of Nodes (or Proteins) in the Network

Biological systems are modular in nature and so are the biological networks. The network is often divided into sets of nodes called as modules or communities such that connectivity of nodes within the module is higher than between the modules [23, 26]. Previous analysis suggested that modules are often found enriched with proteins that are involved in specific biological processes or pathways. Thereby, higher modularity of the network offers partition of various cellular functions. As shown in the Fig. 5.1, the co-expression network is clearly divided into three modules. Each module also consists of proteins that are highly connected with each other than other proteins. Each module consists of proteins that perform related functions. Recently, it was hypothesized that modularization could lead to the enlargement of the network diameter because it increases the minimal path length between modules and because there are usually more pairs of nodes across modules than within modules in a highly modular network [23]. A partition of the large network into subnetworks or modules or communities often ease the analysis. The tools that can be used to partition networks into various modules are Markov cluster algorithm (MCL) which can be accessible from http://micans.org/mcl/, MCODE algorithm [27], and MINE [28].

5.6 Network Visualization and Analysis Tools

Over the years, several tools have been developed to analyze large networks. Most widely used software for visualization of the network are Cytoscape [29], VisANT [30], Pajek (http://vlado.fmf.uni-lj.si/pub/networks/pajek/), NAViGaTOR [31], yED (http://www.yworks.com/en/products_yed_about.html). Cytoscape is one of the best among others since lots of plugins have been developed which can be used in this software for analysis of various aspects associated with biological networks. All the figures that are used in this chapter are generated using Cytoscape. Hive plots is a recently developed package for network visualization [32]. It generates graph layouts, which are quantitative and easy to interpret.

For analysis of topological properties and also for visualization igraph is one of the best packages available to date (http://igraph.sourceforge.net/). It comes with the C libraries and also as a part of CRAN package for R (http://www.r-project.org/). Recently, networks have been studied using organization of the edges instead of nodes [33]. These parameters can be explored by using CRAN package linkcomm [34]. Another good resource for network analysis is network analysis tools (NeAT), where many tools are available for reconstruction, visualization, and statistical analysis of biological networks [35] (http://rsat.bigre.ulb.ac.be/rsat/index_neat.html).

5.7 Scope and Limitations

Although the structural and functional analysis of PPI networks has improved our understanding of the underlying cellular biology, they are not without flaws. It is suspected that what is examined is only a small fraction of entire proteome. Furthermore, what we analyze is just a static picture of the overwhelming complex spatio-temporal interactions that take place in a cell [36]. Nonetheless, the genome-scale protein interaction maps have many practical applications and hold the key in our quest to understand organization of complex biological systems [37].

References

1. Barabasi, A.L., Albert, R.: Emergence of scaling in random networks. Science **286**(5439), 509–512 (1999)
2. Jeong, H., et al.: The large-scale organization of metabolic networks. Nature **407**(6804), 651–654 (2000)
3. Watts, D.J., Strogatz, S.H.: Collective dynamics of 'small-world' networks. Nature **393**(6684), 440–442 (1998)
4. Faith, J.J., et al.: Many microbe microarrays database: uniformly normalized Affymetrix compendia with structured experimental metadata. Nucl. Acids Res. **36**(Database issue), D866–D870 (2008)
5. Barabasi, A.L., Oltvai, Z.N.: Network biology: understanding the cell's functional organization. Nat. Rev. Genet. **5**(2), 101–113 (2004)
6. Jeong, H., et al.: Lethality and centrality in protein networks. Nature **411**(6833), 41–42 (2001)
7. Grigoriev, A.: On the number of protein–protein interactions in the yeast proteome. Nucl. Acids Res. **31**(14), 4157–4161 (2003)
8. Marcotte, E.M., et al.: Detecting protein function and protein–protein interactions from genome sequences. Science **285**(5428), 751–753 (1999)
9. Kim, P.M., et al.: The role of disorder in interaction networks: a structural analysis. Mol. Syst. Biol. **4**, 179 (2008)
10. Borneman, A.R., et al.: Target hub proteins serve as master regulators of development in yeast. Genes Dev. **20**(4), 435–448 (2006)
11. Yu, H., et al.: Genomic analysis of essentiality within protein networks. Trends Genet. **20**(6), 227–231 (2004)
12. Yu, H., et al.: The importance of bottlenecks in protein networks: correlation with gene essentiality and expression dynamics. PLoS Comput. Biol. **3**(4), e59 (2007)
13. Butland, G., et al.: Interaction network containing conserved and essential protein complexes in *Escherichia coli*. Nature **433**(7025), 531–537 (2005)
14. Hu, P.: Global functional atlas of *Escherichia coli* encompassing previously uncharacterized proteins. PLoS Biol. **7**(4), e96 (2009)
15. Butland, G.: eSGA: *E. coli* synthetic genetic array analysis. Nat. Methods **5**(9), 789–795 (2008)
16. Cohen, R., Havlin, S.: Scale-free networks are ultrasmall. Phys. Rev. Lett. **90**(5), 058701 (2003)
17. Chung, F., Lu, L.: The average distances in random graphs with given expected degrees. Proc. Natl. Acad. Sci. USA **99**(25), 15879–15882 (2002)
18. Tanaka, R., Yi, T.M., Doyle, J.: Some protein interaction data do not exhibit power law statistics. FEBS Lett. **579**(23), 5140–5144 (2005)
19. Ivanic, J., Wallqvist, A., Reifman, J.: Probing the extent of randomness in protein interaction networks. PLoS Comput. Biol. **4**(7), e1000114 (2008)
20. Maslov, S., Sneppen, K.: Specificity and stability in topology of protein networks. Science **296**(5569), 910–913 (2002)
21. Hakes, L., et al.: Protein–protein interaction networks and biology—what's the connection? Nat. Biotechnol. **26**(1), 69–72 (2008)
22. Hao, D., Li, C.: The dichotomy in degree correlation of biological networks. PLoS One **6**(12), e28322 (2011)
23. Zhang, Z., Zhang, J.: A big world inside small-world networks. PLoS One **4**(5), e5686 (2009)
24. Latora, V., Marchiori, M.: Efficient behavior of small-world networks. Phys. Rev. Lett. **87**(19), 198701 (2001)
25. Wagner, A., Fell, D.A.: The small world inside large metabolic networks. Proc. Biol. Sci. **268**(1478), 1803–1810 (2001)

26. Ravasz, E., et al.: Hierarchical organization of modularity in metabolic networks. Science **297**(5586), 1551–1555 (2002)
27. Brohee, S., van Helden, J.: Evaluation of clustering algorithms for protein–protein interaction networks. BMC Bioinform. **7**, 488 (2006)
28. Rhrissorrakrai, K., Gunsalus, K.C.: MINE—module identification in networks. BMC Bioinform. **12**, 192 (2011)
29. Shannon, P., et al.: Cytoscape: a software environment for integrated models of biomolecular interaction networks. Genome Res. **13**(11), 2498–2504 (2003)
30. Hu, Z., et al.: VisANT 3.5: multi-scale network visualization, analysis and inference based on the gene ontology. Nucl. Acids Res. **37**(Web Server issue), W115–W121 (2009)
31. Brown, K.R., et al.: NAViGaTOR: network analysis, visualization and graphing Toronto. Bioinformatics **25**(24), 3327–3329 (2009)
32. Krzywinski, M., et al.: Hive plots—rational approach to visualizing networks. Brief Bioinform. br069v1–bbr069 (2011)
33. Ahn, Y.Y., Bagrow, J.P., Lehmann, S.: Link communities reveal multiscale complexity in networks. Nature **466**(7307), 761–764 (2010)
34. Kalinka, A.T., Tomancak, P.: Linkcomm: an R package for the generation, visualization, and analysis of link communities in networks of arbitrary size and type. Bioinformatics **27**(14), 2011–2012 (2011)
35. Brohee, S., et al.: Network analysis tools: from biological networks to clusters and pathways. Nat. Protoc. **3**(10), 1616–1629 (2008)
36. von Mering, C., et al.: Comparative assessment of large-scale data sets of protein–protein interactions. Nature **417**(6887), 399–403 (2002)
37. Raman, K.: Construction and analysis of protein–protein interaction networks. Autom. Exp. **2**(1), 2 (2010)

Chapter 6
Applications of Protein Interaction Networks

Abstract Complex systems are often organized in the form of networks. The last decade has seen numerous breakthroughs in the network theory, which have been implicated in every aspect of biological sciences. In this chapter, we review contributions of the network analysis for understanding cellular organization in system biology era using protein-protein interaction networks.

6.1 Computational Methods Complement Experimental Techniques

Networks provide a simplified overview of the Web of interactions that exist inside a cell. At the end of twentieth century, two groups independently carried out large-scale analysis of physical protein interactions using yeast two-hybrid method in which interactions of protein partners accessed in yeast using a transcriptional readout [1, 2]. Since then protein–protein interaction (PPI) networks have been identified using two-hybrid method in several other model systems such as *Caenorhabditis elegans*, *Drosophila melanogaster*, and humans [3]. More recently, high-throughput studies using affinity purification followed by identification of associated proteins using mass spectrometry resulted in large amount of protein interaction datasets for *Escherichia coli* and Yeast [1, 2, 4–6]. The experimental techniques applied to determine PPI are prone to systematic errors. The compiled PPI data from previous analysis was estimated that more than half of the existing PPI data derived from experimental analysis is likely to be spurious [7]. Moreover, only 2 % overlap is observed among the PPIs predicted using various experimental methods [7]. From the previous analyses, it is also inferred that there is a little overlap between experimentally and computationally identified PPIs. Hence, they can complement each other. The reliability of the available experimental PPI

V. Y. Muley and V. Acharya, *Genome-Wide Prediction and Analysis of Protein–Protein Functional Linkages in Bacteria*, SpringerBriefs in Systems Biology, DOI: 10.1007/978-1-4614-4705-4_6,

datasets can be enhanced by filtering out spurious interactions using computational predictions. Considering the vast number of pairs among proteins of any organism, it is not only expensive but also current technology limits experimental test for even a subset of these interactions. Furthermore, high-throughput analysis could be performed only in the standard laboratory conditions; the current set of experiments may not identify interactions for proteins that conditionally express or interact with each other.

It has shown that predictions at genomic context level actually had both a 7.7 % higher coverage and 5.3 % higher accuracy than mRNA co-expression (expression similarity) [7, 8]. It has also shown true for direct experimental techniques such as yeast two-hybrid analysis or high-throughput mass spectrometric protein-complex identification [7, 8].

6.2 Network-Based Protein Function Predictions

The percentage of uncharacterized proteins even in model organisms is quite high. The positions of these proteins in a network play an important role in elucidating their cellular functions. There are two major approaches applied in the context of network to predict protein functions.

6.2.1 Neighborhood or Guilt-By-Association-Based Function Predictions

The simplest and the most direct method determines the function of a protein based on the known function of proteins lying in its immediate neighborhood. [9, 10]. Schwikowski et al. [9] have predicted three functions for a given protein that are most common among its neighbors. It is a simple and effective approach but does not consider full topology of network and assign significant values for predictions. Hishigaki et al. [11] tried to tackle the first problem by computing objective function for scoring functional assignments. The functions for a protein were predicted by detailed examination of its adjacent neighbors as well as their neighborhood in a network.

6.2.2 Module-Based Function Predictions

The second scheme identifies several modules/communities in the network, each represents a set of proteins highly connected with each other than with rest of the proteins in the network. Each module can then be assigned probable functions based on the proteins with known function therein. This approach is called as module-assisted annotation [12–14].

6.3 Dynamic Analysis of Biological Processes

Computational analyses of predicted networks have immense potential in aiding our understanding about gene/protein function, biological pathways, and cellular organization [15, 16]. However, the functioning of cells and organisms is mainly regulated by dynamic interactions. The predicted PPI networks along with other high-throughput experimental datasets such as gene expression allow us to study biological systems at different times and conditions. These studies are increasingly helpful in elucidating interaction dynamics and emerging as a new subfield within the computational biology [17]. Several studies on integration of PPI networks and gene expression data promise to be a major step forward in our ability to model and reason about cellular function and behavior [18, 19].

6.4 Cross-Talk Among Cellular Pathways

Over the years significant amount of data have been generated for interactions among cellular components. These data include interactions gathered through individual studies, large-scale screens, and also have been assembled from the literature into various databases. Proteins are center of attraction among the other cellular components due to their indispensible role in almost every biological process that occurs in the cell. Thus, most of the networks reconstructed so far have been focused on proteins. To date, at least five types of biological networks have characterized in detail. These networks are genetic interaction, PPI, transcriptional regulatory interaction, metabolic interaction, and signaling interaction (i.e., protein phosphorylation) [15, 20–23]. In addition, drugs and their targets are also represented as networks and studied in detail but only in eukaryotes [24, 25]. Signaling network is one of the promising areas of research due to their complex organization. Furthermore, integrative analysis of metabolic, signaling, and transcriptional regulatory networks can be used to understand the cross-talk between their components.

References

1. Ito, T., et al.: A comprehensive two-hybrid analysis to explore the yeast protein interactome. Proc. Natl. Acad. Sci. USA **98**(8), 4569–4574 (2001)
2. Ito, T., et al.: Toward a protein–protein interaction map of the budding yeast: a comprehensive system to examine two-hybrid interactions in all possible combinations between the yeast proteins. Proc. Natl. Acad. Sci. USA **97**(3), 1143–1147 (2000)

3. Li, D., et al.: Protein interaction networks of *Saccharomyces cerevisiae*, *Caenorhabditis elegans* and *Drosophila melanogaster*: large-scale organization and robustness. Proteomics **6**(2), 456–461 (2006)
4. Butland, G., et al.: Interaction network containing conserved and essential protein complexes in *Escherichia coli*. Nature **433**(7025), 531–537 (2005)
5. Hu, P., et al.: Global functional atlas of *Escherichia coli* encompassing previously uncharacterized proteins. PLoS Biol. **7**(4), e96 (2009)
6. Zeghouf, M., et al.: Sequential peptide affinity (SPA) system for the identification of mammalian and bacterial protein complexes. J. Proteome Res. **3**(3), 463–468 (2004)
7. von Mering, C., et al.: Comparative assessment of large-scale data sets of protein–protein interactions. Nature **417**(6887), 399–403 (2002)
8. Huynen, M.A., et al.: Function prediction and protein networks. Curr. Opin. Cell Biol. **15**(2), 191–198 (2003)
9. Schwikowski, B., Uetz, P., Fields, S.: A network of protein–protein interactions in yeast. Nat. Biotechnol. **18**(12), 1257–1261 (2000)
10. Aravind, L.: Guilt by association: contextual information in genome analysis. Genome Res. **10**(8), 1074–1077 (2000)
11. Hishigaki, H., et al.: Assessment of prediction accuracy of protein function from protein–protein interaction data. Yeast **18**(6), 523–531 (2001)
12. Song, J., Singh, M.: How and when should interactome-derived clusters be used to predict functional modules and protein function? Bioinformatics **25**(23), 3143–3150 (2009)
13. Sharan, R., Ulitsky, I., Shamir, R.: Network-based prediction of protein function. Mol. Syst. Biol. **3**, 88 (2007)
14. Voevodski, K., Teng, S.H., Xia, Y.: Finding local communities in protein networks. BMC Bioinform. **10**, 297 (2009)
15. Zhu, X., Gerstein, M., Snyder, M.: Getting connected: analysis and principles of biological networks. Genes Dev. **21**(9), 1010–1024 (2007)
16. Aittokallio, T., Schwikowski, B.: Graph-based methods for analysing networks in cell biology. Brief Bioinform. **7**(3), 243–255 (2006)
17. Przytycka, T.M., Singh, M., Slonim, D.K.: Toward the dynamic interactome: it's about time. Brief Bioinform. **11**(1), 15–29 (2010)
18. de Lichtenberg, U., et al.: Dynamic complex formation during the yeast cell cycle. Science **307**(5710), 724–727 (2005)
19. Luscombe, N.M., et al.: Genomic analysis of regulatory network dynamics reveals large topological changes. Nature **431**(7006), 308–312 (2004)
20. Babu, M.M., et al.: Structure and evolution of transcriptional regulatory networks. Curr. Opin. Struct. Biol. **14**(3), 283–291 (2004)
21. Hyduke, D.R, Palsson, B.O.: Towards genome-scale signalling-network reconstructions. Nat Rev Genet. **11**(4), 297–307 (2010)
22. Yamada, T., Bork, P.: Evolution of biomolecular networks: lessons from metabolic and protein interactions. Nat. Rev. Mol. Cell Biol. **10**(11), 791–803 (2009)
23. Ravasz, E., et al.: Hierarchical organization of modularity in metabolic networks. Science **297**(5586), 1551–1555 (2002)
24. Zhu, M., et al.: The analysis of the drug-targets based on the topological properties in the human protein–protein interaction network. J. Drug Target. **17**(7), 524–532 (2009)
25. Yildirim, M.A., et al.: Drug-target network. Nat. Biotechnol. **25**(10), 1119–1126 (2007)